URBAN FORESTRY

URBAN FORESTRY

SECOND EDITION

GENE W. GREY
Society of American Foresters

FREDERICK J. DENEKE
(United States Department of Agriculture)
Extension Service

KRIEGER PUBLISHING COMPANY
MALABAR, FLORIDA
1992

Original Edition 1986
Reprint Edition 1992

Printed and Published by
KRIEGER PUBLISHING COMPANY
KRIEGER DRIVE
MALABAR, FLORIDA 32950

8194 BdT 45.00

Library of Congress Cataloging-In-Publication Data
Grey, Gene W., 1931-
    Urban forestry / Gene W. Grey, Frederick J. Deneke. -- 2nd ed.,
  Reprint ed.
      p.   cm.
  Reprint. Originally published: New York : Wiley, c1986.
  Includes index.
  ISBN 0-89464-704-0
  1. Urban forestry. I. Deneke, Frederick J., 1942-
II. Title.
[SB436.G73 1992]
635.9'77'091732--dc20   91-46194
                                   CIP

10   9   8   7   6   5   4   3   2

TO HAROLD GALLAHER,
WHO MADE IT POSSIBLE

FOREWORD

The urban forests of America are great green resources enhancing city life, while often battling the elements of city life for survival. From the air, many cities seem to be nestled between trees, but from the ground, the picture is different. The forest consists of clumps and patches of green separated by both concrete and claims of ownership. Relatively few recognize this forest as forest, and even fewer recognize its need for comprehensive management. The urban forester is a trustee of this valuable ecological resource, and its management requires special knowledge and skills. Understanding the plant kingdom is but a part of the forester's work; understanding people and the institutions they have constructed presents the larger challenge.

For many years foresters were not intensively involved with the landscapes of cities. During the last 20 years that picture has changed. The forestry profession is examining urban forestry and is tailoring management techniques to these areas. The skills of recognizing and managing a natural system of plants is the foundation of the forestry profession. Yet, the profession is learning from other disciplines skilled in plant materials, design, recreation, and plant culture—thus it is merging the knowledge of forestry with the skills of other professionals.

The scope of the management job in urban areas is large. There are approximately 69 million acres of urban and suburban land in this country (this number does not include the thousands of smaller rural communities). Trees and other plants that set their roots in populated areas must often survive under difficult conditions. They must endure limited planting spaces and poor soils while competing with automobiles, lawnmowers, power lines, and many other utilities. The urban forest manager must meet the challenge of addressing these physical factors while fitting management strategies into the framework of the local social and political system. To keep the trees healthy in the urban forest requires considerable planting of seeds in the minds of community leaders, and cultivation within the walls of government offices.

The opportunity to improve the ecology of cities starts before the first house is planned or the first road is built at the city's edge. Designing a community to fit into the landscape is more than just a dream of Ian McHarg who wrote *Design with Nature*. Foresters have an opportunity to build on McHarg's concept and to apply their knowledge of forest ecology to the land planning process. These land-use decisions should not be delegated to any single profession. Foresters have been noticeably absent from this planning process in the past, and the cities have suffered from their absence. Existing natural features, soils, and vegetation are valuable resources that should be planned into developing cities and towns. Many environmental problems with trees in urban areas can often be traced to stress resulting from soil conditions, limited growing space, pollution, and wounding. Many of these problems could be avoided by designing with nature.

Today, the urban forestry professional has a vast amount of knowledge from which to draw. These data have been collected both in the field and in the laboratories of researchers. The profession is growing and the workplace is changing.

The authors of this book actively pursued the science of urban forestry during its developing years, and reflect these changes in this second edition of *Urban Forestry*. The science of urban forestry has its roots in the heart of this nation's ecological wealth. The practice of urban forestry combines the science with the near-infinite complexity of social, economic, and political factors of urban life.

The forests and trees of our urban areas are essential elements of a community; essential to both the environment and to the people who spend their lives in the community. Urban foresters have much to contribute to the quality of life in the communities of tomorrow. This book helps supply the tools needed for the urban forester to manage effectively this special forest.

Gary Moll
The American Forestry Association

Washington D.C.

PREFACE

In our original text, written in 1977, we stated that "most cities are indeed forests—communities of trees and other woody plants both on and in the soil. One has only to stand on a high vantage point or fly over a city to perceive that below is a true forest." We also wrote that "although the benefits of trees and other vegetation in and near cities and towns have long been recognized, only recently has the concept of urban trees as a forest generally been recognized and the proper management systems applied." We stand firm on our first statement, that most cities are indeed forests. We confess, however, that we may have been premature in our assertions that the concept of urban trees as a forest has generally been recognized and that proper management systems are being applied. Wisdom gained through the years since would now have us write that the concept of urban trees as a forest is slowly gaining recognition, and that management systems based on this recognition are being applied in some instances.

To solidify the point that cities are forests—and this point is fundamental to urban forestry—we stated that the urban forest is composed of trees along streets, in parks, around public buildings, and on all kinds of other public and private property. It is located on all land use zones—residential, recreational, business, and industrial. This forest is in a near-constant state of adjustment to the use of the land on and in which it grows. Construction, repair, and demolition of streets, sidewalks, waterlines, sewers, power and communication lines, buildings, parking lots, and other structures occur continually somewhere in the forest.

Management of the urban forest is the responsibility of public and private owners. It involves any combination of property owners, city park or forestry departments, city tree boards or commissions, private tree care firms, nursery people, and others. Optimum management, however, requires a system that considers both the needs of individual trees and the forest as a whole. Ideally, it must be a system that provides for the social values of the forest while protecting the rights of all property owners as much as possible.

It is to this system, immature as it is, and the various elements that relate to it, that we address this book. It is our hope in so doing that we might advance urban forestry to make accurate our original statement that the concept of urban trees as a forest is generally recognized and that proper management systems are applied.

Although the boundaries of urban forestry are still sometimes disputed, there is no denying its presence. It has advanced from an emerging concept—not welcomed by all—in the late 1960s and early 1970s to a full-blown movement in the 1980s. Traditional foresters are a bit uncomfortable with it. Arborists, although not as vocal as before, view it as somewhat of a threat. The USDA Forest Service does not know quite what to do with it. But all know it is there, and that it deals with forests where people live, and may thus be of considerable importance.

A driving force behind urban forestry, hoewever, is the conviction of some that it is the bridge to forestry's future. This conviction was expressed by Henry J. Vaux in a May 1980 *Journal of Forestry* article:

> *The recent emergence of urban forestry as a focus of substantial interest is one of the most important developments in American forestry in a long time. Its importance is not just for foresters who practice in urban areas, but for all foresters, wherever they work.*
>
> *From my perspective, urban forestry, far from being tangential or even extraneous to real dirt forestry (as some of us seem to feel), actually lies at the very heart of our profession's contemporary problems. It offers us a bridge to the people who ultimately decide what we can and cannot do. How we use that bridge probably will determine as much as anything else, our professional future.*
>
> *We can choose to use the bridge to try to persuade these strange urbanites that we know what is best—that our scientific and practical knowledge permits us to know and understand urban needs. We can try that course, but it may well prove a tragically lost opportunity for the profession. We must avoid a missionary approach.*
>
> *But if we use the bridge as a new and significant way of gaining insight, knowledge, and personel relationships in a sector of society from which we have too long been aloof, urban forestry can provide us with a means to solve some deep-rooted problems, both inside and outside the urban area. Let us take the approach of the student—of students who come to learn for their own benefit as much as to benefit those for whom they work. Foresters who work in urban areas understand the need for this approach. But others clearly do not. Let us work to create that understanding.*

Perhaps this is but another way of saying that although a strong movement, urban forestry has not yet matured. Let us proceed, with the hope as stated previously, that we might help hasten its maturity.

Gene W. Grey
Frederick J. Deneke

ACKNOWLEDGMENTS

We are particularly grateful to the students, urban forestry practitione͚ and others who found the first edition of *Urban Forestry* a worthwhile resu͙ of our efforts; and to our wives, Sandra and Jody, for their encouragement and patience. We thank also those upon whose work we have drawn in this effort, and especially all those who have brought urban forestry from dawn into the full sunlight of reality.

G.W.G.
F.J.D.

CONTENTS

Chapter 1   **History of Urban Forestry**   **2**

Early History   2
North America   3
Recent Developments   5

Chapter 2   **Distribution and Ownership of the Urban Forest**   **14**

Public Lands   17
Private Lands   24
Public or Private Management Responsibility   25

Chapter 3   **Composition of the Urban Forest**   **30**

Purpose or Function   32
Popular Species   32
Public Control   33
Socioeconomic Factors   33
Mobility   34
Nostalgia   35
Other Factors   35
Manhattan, Kansas   36

Chapter 4   **Benefits of the Urban Forest**   **50**

Climate Amelioration   50
Engineering Uses   65
Architectural Uses   95
Economic Benefits   102
Esthetic Uses   103
Wildlife Benefits   103
Urban Wildlife Planning   109
Other Uses   112

Chapter 5   Environment of the Urban Forest   118

Physical Environment   118
Space   120
Soils   122
Topography   123
Microclimate   124
Pollution   128
People   133

Chapter 6   Management of the Urban Forest   138

Classifying and Inventorying the Urban Forest   140
Urban Forest Inventories   141
Planting   144
Maintenance   160
Contracts   170
Removal   174

Chapter 7   Monetary Values of the Urban Forest   184

Alternative Values   184
City Assets   184
Maintenance Values   187
Timber Values   187
Property Values   192
Legal Values   193
Evaluation Formulas   202

Chapter 8   Municipal Forest Administration   208

Political and Policy Environment   208
Administrative Structure   209
Ordinances   211
Financing   215
Urban Planning and Urban Forestry   217

**Chapter 9   Information, Education and
                Training**   **226**

    Information and Education   226
    Training   230

**Chapter 10 Urban Forestry Programs, Support
                Organizations and Research**   **242**

    State Forestry Programs   242
    Consulting Urban Foresters   245
    Support Organizations   246
    Research   252

Chapter 11 Urban Forestry Issues   **258**

APPENDIX 1   Schools for Urban Forestry in North
                  America   262
APPENDIX 1a   Sample Urban Forestry Curriculum
                  (University of Minnesota)   264
APPENDIX 2   Sample City Tree Ordinance for a Small
                  Midwestern Community   265
APPENDIX 2a   Model Municipal Tree Ordinance for the
                  Atlanta, Georgia Area   270
APPENDIX 2b   Aboricultural Specifications and Standards
                  of Practice   282
APPENDIX 3   A City Street Tree Inventory System   288
APPENDIX 4   Sample Format for a Call for Bids   291
APPENDIX 5   Conversion Factors   294

Index   295

1
HISTORY
OF URBAN
FORESTRY

Although the term "urban forestry" is relatively new (Jorgensen, 1970), certain practices and disciplines that make up the field of urban forestry have been long established. The use of trees to enhance the environment and many of the principles relating to their care are extremely old. Indeed, one can find biblical reference to the planting of trees in the Garden of Eden.

Early History

Trees have been esthetically important to people since earliest civilization. The Egyptians, Phoenicians, Persians, Greeks, Chinese and Romans held trees in high esteem and in certain situations worshipped them. They used trees for their esthetic benefits, developing formal gardens and sacred groves to enhance temple settings. Even in these early cultures trees were also used to complement statues and provide a landscape for buildings. Along with these uses developed a rudimentary knowledge of tree care. Transplanting of trees was common as early as 1500 B.C. in Egypt (Winters, 1974). Theophrastus and Pliny both contributed much knowledge in these early times toward the maintenance and care of trees (Chadwick, 1970).

This knowledge continued to develop as civilization advanced. Botanical gardens began to evolve during the Middle Ages with particular emphasis on plants with medicinal properties. Since this was a time of separatism and survival, such gardens were necessary features. As the Renaissance period evolved, man embarked upon new adventures in scientific achievement and trade. As part of this increased trade and travel plants from other countries were introduced. These plant species were exhibited in private gardens and collections and led to the establishment of large botanic gardens in many countries. This increase in plant species resulted in greater knowledge of plants and their care.

Early England contributed much to our knowledge concerning trees. The first recording of the term "arborist" can be found in James Lyte's book *Dodens* in 1578 (Chadwick, 1970). *A New Orchard and Garden,* a book written by William Lawson in 1618, is quite detailed in its references to tree maintenance as we know it today (Chadwick, 1970). John Evelyn in 1662 wrote a monumental work on trees called *Sylva* (Evelyn, 1776). In it he discussed all known aspects of forestry and fruit trees and included references to past works on trees. In Evelyn's book there are passages concerned with arbor walks and the ornamental use of trees along with his recommendations on which species to plant. There are many other early books on trees, but suffice it to say that by the 1700s the use of trees in urban settings, a respect for their importance in natural areas, and the knowledge of their growth and maintenance were well developed. Botanic gardens had ceased to be as important for medicinal purposes and became places to study horticultural and arboricultural practices. Interest in landscape design was developing, and visitors to garden areas came away with new ideas for their own homes. Thus the use of plants in ornamental settings became more widespread. As time passed, plant materials became more available and designs became less formal as the emphasis shifted from decoration to relaxation and enjoyment.

In the early 1800s the open spaces in various London squares were laid out with freely planted trees and lawns. In fact the image of London at that time was one of trees and grass. Many of these were residential green spaces surrounded by housing, resulting in quiet, comfortable residential neighborhoods (Zube, 1973). Tree-lined boulevards were introduced to Paris in the mid-1800s. The main reason for boulevards, however, was to control movement of troops, and the trees provided them with some measure of defense (Zube, 1973). Such plantings also provided beauty and unified architectural elements.

North America

In early settlement years, trees were often considered a detriment. Though providing fuel, shade, and food, they harbored real and imagined dangers ("horror sylvanum") and occupied soils desirable for food crop production. The popular image of a New England town square with trees and grass did not evolve until the late 1700s and early 1800s (Zube, 1973). Philadelphia did not have street trees until the late 1700s because prior to that time insurance companies would not insure houses that had trees in front of them (Zube, 1973). Eventually cities developed and expanded into forested areas, particularly in eastern North America. The forested areas were modified considerably. Demands would often dictate their total removal; in other areas some native trees remained in the home setting; and on occasion certain forested areas that posed site difficulties in construction would be left intact and development would evolve around them. Many of these forested areas eventually would become parks. In still other areas, such as the Great Plains, urban forests were to be created from areas naturally devoid of woody vegetation.

Settlers and immigrants often brought seed or propagation materials of their favorite tree species from their homelands. These found their way into the "landscape" around homes in both urban and rural settings. In areas cleared or devoid of forests, they were our first street trees. Botanic gardens were occasionally developed by private individuals. Often the decision to incorporate a given plant species was because of its rarity or botanic

3

interest. These gardens were the proving grounds for exotic plant species that would eventually find their way into the general urban landscape. Examples of some of the early botanic gardens were: Henry Shaw's collection (now the Missouri Botanical Garden in St. Louis), the James Arnold Arboretum in Jamaica Plains, Massachusetts, and Longwood Arboretum in Kennett Square, Pennsylvania. During the late 1700s and 1800s several tree species were introduced that became popular lawn and street trees; these were the Lombardy poplar, Norway maple, English elm, and ailanthus (tree-of-heaven). The planting of introduced tree species in lawns and gardens was quite popular until the mid-1800s. This happened primarily for two reasons: (1) most of the early trained horticulturists and foresters were of European origin and training and favored trees with which they were familiar; and (2) most of the early nurseries obtained their ornamental stock from Europe. It should be noted that planting European Stock persisted even until the 1900s, and often the planting stock of native North American tree species originated from Europe. This presented other problems, however, such as the introduction of exotic insect and disease pests. White pine blister rust on eastern white pine originated in such a manner.

As the use of plants in lawn and street settings became more widespread, so did the science of caring for them. Horticultural societies were organized and numerous books and magazines about gardening were published. In the mid-1800s, there developed more of an impetus to use native instead of exotic species. One of the leaders in this movement was Andrew Jackson Downing. He worked with Calvert Vaux and Frederick Olmstead in designing parks and wrote several books on landscaping. His principle was to select the best indigenous trees, believing that the soil and climate of the area would most favor these. He was especially fond of the native maples, oaks, and elms for city tree plantings. His influence can still be found in many cities established after 1850. Of course one cannot leave this era without noting the influence of Frederick Olmstead on city park development. The classic example of his work is Central Park in New York City. Anyone who has lived in or visited New York can appreciate his brilliance and foresight. His work is evidenced in Chicago and other cities and has been emulated in countless others.

The landscape concept in the United States grew as a result of increasing industrialization and was used in the development of suburbs during the mid-1800s. Suburbs were designed to allow an escape to areas of natural beauty; streets curved through wooded and rolling hills. Llewellyn Park in West Orange, New Jersey, is one of the earliest of these towns. If trees were not already present, they were added to give a natural woodland effect (Zube, 1973).

In 1872, J. Sterling Morton, a farmer–legislator, then living in a rela-

4

tively treeless Nebraska, proposed that an annual Arbor Day be observed especially for the purpose of planting trees—a practice that is now observed in most of North America.

It should be noted that one of the early texts of the twentieth century dealing with street tree maintenance and arboriculture practice was issued by a forester of great renown. B. E. Fernow published a book in 1911 titled *The Care of Trees in Lawn, Street and Park.* At the time he was on the faculty of the department of forestry at the University of Toronto, Canada. Jorgensen comments that this was the birthplace of both forestry education and arboriculture in Canada (Jorgensen, 1970). This is probably also true for urban forestry in North America; in fact he devoted one chapter in his book to "esthetic forestry." However, Fernow disagreed that the term "forestry" or "forester" should be applied to street and lawn trees. He preferred the term "tree warden." The term proposed by Fernow was not new. Towns in massachusetts and other northeastern states had tree wardens as early as the 1700s.

The practice of arboriculture also took some giant steps forward in the early twentieth century. At the turn of the century, John Davey, often called the father of modern arboriculture, founded a company that specialized in tree maintenance (Wysong, 1972). In 1924, the International Shade Tree Conference (now International Society of Arboriculture) had its beginning. Here people interested in the care and management of shade trees could meet, share ideas, and further advance the field of arboriculture. The Society marks the first organizational effort by both lay and professional people to respond to public concern for the planting and care of lawn and street trees.

In the 1930s the National Park Service approved the special care of trees in areas of concentrated use or of historical significance within their jurisdiction. A. Robert Thompson, a forester in the park Service, was commissioned to write a series of bulletins on tree maintenance. The Civilian Conservation Corps (CCC) also made a significant contribution by its involvement not only in afforestation, reforestation, and windbreak plantings, but also in the planting of street trees in many cities.

Recent Developments

Past developments focused on tree planting, tree maintenance, and landscape architecture. The concept of urban forestry or total management of the urban forest system did not develop until the mid-1960s. However, pressures toward building such a management concept came as early as the 1930s. Ironically, the major impetus for such a concept was the result of factors outside the developing system. Dutch elm disease, phloem necrosis, and oak wilt were among them. With the advance of such diseases came a recognition for the need of knowledge and management systems to cope with them. Research was directed to appropriate areas and specific courses in arboriculture were introduced at many universities. The need for individuals competent in shade tree management was recognized in many cities. Individuals, and in many cases entire staffs, were added to cope with growing problems of managing the tree resources in urban areas. Individually these became, for example, the city forester, the tree warden, and the municipal arborist. Collectively, they became the park and tree departments or the park and landscape divisions.

Yet such programs were still limited in scope and focused on the concept of the individual tree and its needs. Cities had not yet dealt with the concept of integrated urban forest ecosystem management. The concept of urban forestry was introduced first at the University of Toronto in 1965 (Jorgensen, 1970). Jorgensen states that urban forestry as developed in Canada

> *does not deal entirely with city trees or with single tree management, but rather with tree management in* the entire area *influenced by and utilized by the urban population. This area naturally includes the watershed areas and the recreational areas serving the urban population, as well as the areas lying between these service areas and politically designated urban areas and its trees. The politically established boundaries for municipalities rarely include the entire geographical area influenced by urbanization.*

An early spokesman for urban forestry in the United States was Professor John W. Andresen, formerly Forestry Chairman at Southern Illinois University, Carbondale. Professor Andresen in now Director, Urban Forestry Studies Programme at the University of Toronto.

In 1967 a report was published by the Commission on Education in Agriculture and Natural Resources that referred to the need for foresters to be responsive and sympathetic to an increasingly urban America (Commission on Education in Agriculture and Natural Resources, 1967). The rationale was that the urban trend would continue and that foresters would be forced to respond to the demands of growing urban constituencies. In fact a 1972 Louis Harris poll, a section of which involved desired American life-styles and environmental values, reported that 95 percent of those polled listed "green grass and trees around me" as an important environmental value (Zube, 1973).

In 1968, the Citizens Advisory Committee on Recreation and Natural Beauty, chaired by Laurance S. Rockefeller, submitted their Second Annual Report to the President of the United States. Part of the report focused on the theme that city trees constituted a resource that was not being adequately cared for. They recommended that

> *an urban and community forestry program be created in the United States Forest Service. The program should encourage research into the problems of city trees, provide financial and technical assistance for the establishment and management of city trees and develop Federal training programs for the care of city trees.*

It was also suggested that

> *The U.S. Forest Service should create an urban and community forestry program in cooperation with the states to protect, improve, and establish trees in community, suburban and urban areas. A Federal–State program would provide technical and financial assistance to local governments, organizations and individuals to establish and manage trees and related plants in community parks, open spaces, streets, greenbelts, and on private properties.*

With the acceptance of this report by the President came the official recognition of urban forestry in the United States.

In response to this challenge, the Pinchot Institute of Environmental Forestry Studies was created in 1970. This is an interdisciplinary division of the USDA Forest Service, Northeastern Forest Experiment Station in cooperation with several universities. Its express purpose is to improve, through environmental forestry research, human environments in the densely populated areas of the Northeast. In October 1971, U.S. Congressman Robert L. F. Sikes (Florida) introduced a bill, H.R. 8817, that included authorization for the expenditure of $5 million for a new program of urban environmental forestry. Although this bill was passed by both houses of Congress, the President vetoed it in deference to other national priorities.

In 1972, the Cooperative Forest Management Act of 1950 was ammended by Public Law 92-288. This gave the U.S. Forest Service responsibility for developing an active program in urban forestry. This program was funded for fiscal year 1978 at $2.5 million. Administered by the Forest Service's division of State and Private Forestry, these funds allowed or strengthened state forestry department programs to provide urban forestry technical assistance to local governments, organizations, and individuals. Solid programs were built in several states. Notable among these were Florida, Georgia, Maryland, Missouri, Kansas, Colorado, and California. Programs in Georgia and Florida were concentrated on large metropolitan areas. Kansas and Colorado emphasized smaller towns. Maryland worked largely with urban

7

expansion into natural forests. California's approach, in large measure, was to facilitate and enhance the efforts of existing organizations, particularly volunteers. The California program was expanded greatly by passage of AB–320 in 1979, which designated ten percent of receipts from timber sales on State Forests for urban forestry.

Recent efforts to develop urban forestry programs have not been limited to governmental agencies. In 1972, the Society of American Foresters initiated an Urban Forestry Working Group. The Working Group put forth this policy statement:

> *Urban forestry is a specialized branch of forestry that has as its objective the cultivation and management of trees for their present and potential contribution to the physiological, sociological and economic well-being of urban society. Inherent in this function is a comprehensive program designed to educated the urban populace on the role of trees and related plants in the urban environment. In its broadest sense, urban forestry embraces a multi-managerial system that includes municipal watersheds. wildlife habitats, outdoor recreation opportunities, landscape design, recycling of municipal wastes, tree care in general, and the future production of wood fiber as raw material.*

Activity within the Society of American Foresters continued. In 1980, there were more than 700 members in the Urban Forestry Working Group, making it the thirteenth largest of all 29 working groups. In 1980, the Forest Sciences Board of the SAF, reflecting the uneasiness of the general membership about urban forestry, requested a "White Paper" to explain urban forestry as perceived and practiced, and to provide a base for SAF policies and activities. This paper (Grey et al.,) recommended that SAF:

Recognize and support urban forestry as a vital part of resource management. This support should be shown in the *Journal of Forestry*, at SAF meetings, and in relations with federal, state, and local governments.

Assist urban foresters in educating people about urban forest management.

Recognize that professionals other than foresters are involved, and establish and maintain communication with these individuals and their professional societies.

Monitor and suggest improvements in curricula so that urban forestry education will adequately respond to the needs of modern society.

Recognize that urban people exert a strong influence on natural resource issues in rural areas as well as urban areas, and urban forestry provides an opportunity to reach these people.

Allocate time during national conventions for urban forestry topics.

Continue to encourage articles and reports on urban forestry in the *Journal of Forestry*.

The International Society of Arboriculture (formally the International Shade Tree Conference) formed an Urban Forestry Committee in 1973. The ISA's official publication, the *Journal of Arboriculture*, has been an important forum for articles relating to urban forestry.

An event of great significance in urban forestry was the 1978 First National Urban Forestry Conference. Held in Washington, D.C., and sponsored by the USDA Forest Service and the State University of New York College of Environmental Science and Forestry, this landmark conference brought together over 500 individuals representing all aspects and interests in urban forestry. Its objectives were twofold: to present the "state of the art" of urban forestry and to facilitate better understanding and cooperation between individuals and interests. The conference was divided into three general sessions: (1) the social, economic, and physical benefits from urban forests; (2) culture and protection activities of urban forests; and (3) the planning and managing of urban forests. It is felt (Herrington, 1979) "that the conference did provide for a convergence of views and a melding of ideas in the field." This statement is from the proceedings of the conference published in 1979. The proceedings represent an overview of the scope of urban forestry at that time.

Co-sponsorship of the conference by the State University of New York College of Environmental Science and Forestry was largely a product of their role in the Pinchot Institute of Environmental Forestry. Organized in 1970 as a consortium of ten northeastern universities, the Pinchot Institute addressed the urban forestry research needs of that area. The Institute was supported by the USDA Forest Service and coordinated by the Northeastern Forest Experiment Station at Upper Darby, Pennsylvania. The university researchers were organized as the Consortium for Environmental Forestry Studies. In 1982, the Institute was renamed and its support by the USDA Forest Service substantially reduced.

In the early 1960s, the USDA Forest Service acquired Grey Towers, the former home of Gifford Pinchot in Milford, Pennsylvania. This facility has since been used as the headquarters for the Pinchot Institute for Conservation Studies, with the purpose of furthering conservation programs of the Forest Service and other conservation agencies through research, training, and conferences. In 1977, the program was expanded to include urban forestry with the objective of improving programs through the organization of national information sharing systems, national and regional conferences, field studies, and publications. This program was also curtailed in 1983, with urban forestry being eliminated.

9

The curtailment of both of the above programs is a manifestation of current Forest Service policy concerning urban forestry. This policy is born of the dilemma of the recognition of urban forestry's potential—its contribution to enhanced benefits for urban people and its educational influence on all of forestry—and the reality of reduced federal budgets, driven to a large degree by forest industries who see little direct corporate benefits from urban forestry. In an attempt to assess the role of the Forest Service in urban forestry in its State and Private Forestry division in this environment, one recent analysis (Millen and Randall, 1981) recommended several involvement categories that could coincide with various appropriations levels. In decreasing order of priority these were: (1) technology transfer; (2) promotion; (3) limited advice or assistance; (4) limited training activities for practitioners at regional levels; (5) limited co-sponsorship of nationally significant demonstration projects; and (6) grant funds to states. Whether this was written as a self-fulfilling prophecy is a matter of conjecture, but events both before and since suggest its adoption.

Seeking a practical way to fulfill a less than unanimous obligation to urban forestry in the face of diminishing program budgets, personnel of the Forest Service and The American Forestry Association met in August of 1980 and agreed that The American Forestry Association "should mount a major effort to bring urban forestry to the front in urban planning and programming through an intensive contact and pin-pointed educational effort to reach, inform and influence the top level policy makers and authorizers in urban affairs" (Gray, 1980). It was also agreed that AFA would seek a major grant to hire a qualified person and conduct a strong effort in urban forestry. The major grant was later to come from the Forest Service when in 1983, adequate funds were transferred to AFA for a full-time urban forestry position.

The American Forestry Association's commitment to urban forestry was manifested in 1981 by the formation of the National Urban and Community Forestry Leaders Council. This organization, described in detail in Chapter 10, was developed by a core group of urban forestry leaders from USDA Extension and Forest Services, Maryland State Forest Service, The American Forestry Association, and American Forestry Institute. Its primary purpose is to provide a unified national support base for urban and community forestry. The inclusion of the word "community" in the title is a reflection of concern by Forest Service administrators that the words "urban" and "forestry" may be viewed by some as contradictory, and thus by calling it urban and community forestry greater appeal to rural audiences, particularly legislators, may result.

An early expression of the USDA–AFA partnership in urban forestry was the Second National Urban Forestry Conference in Cincinnati, Ohio in

October 1982. In addition to AFA and USDA Forest Service and Extension Service, the conference was co-sponsored by the Ohio Forestry Association, the Ohio Department of Natural Resources, and the Cincinnati Park Board. It was held in conjunction with AFA's 107th annual meeting and also commemorated the centennial of the First American Forest Conference held in 1882 in Cincinnait. The conference brought together the most knowledgable assemblage of urban forestry experts ever convened (Rooney, 1983). Because of its union with AFA's annual meeting, it also exposed urban forestry and its vital potential to those who exert a strong influence on all of forestry in the nation. The emphasis of the conference was less technical than practical and concentrated largely on political and administrative methods and program delivery systems. Perhaps the strongest aspect of the conference was its exemplification of the vitality of urban forestry, expressed as a belief and commitment of those present. As Rexford Resler, Executive Vice President of AFA, said of urban forestry in the words of Voltaire: "An idea whose time has come is stronger than all the armies in the world."

BIBLIOGRAPHY

Andresen, J.W., *Community and Urban Forestry: A Selected and Annotated Bibliography*, USDA, Southeastern Area State and Private Forestry, 195 p., 1974.

Andresen, J.W., and B. M. Williams, "Urban Forestry Education in North America," *Journal of Forestry,* 73(12):786–790, 1975.

Chadwick, L.C., "3000 Years of Arboriculture—Past, Present, and Future," in *Proceedings of the 46th International Shade Tree Conference,* 73a–78a, 1970.

Citizens Advisory Committee on Recreation and Natural Beauty, *Second Annual Report to the President,* 1968.

Commission on Education in Agriculture and Natural Resources, *Undergraduate Education in the Biological Sciences for Students in Agriculture and Natural Resources,* Publ. 1495, National Academy of Sciences, Washington, D.C., 1967.

Fernow, B.E., *The Care of Trees in Lawn, Street and Park,* H. Holt and Co., New York, 392 p., 1911.

Gray, J., correspondence to Rexford A. Resler, files of Pinchot Institute for Conservation Studies, Milford, Pa., August 11, 1980.

Herrington, L., "Preface," in *Proceedings of the National Urban Forestry Conference,* Volume I, ESF Publication 80–003, SUNY, Syracuse, N.Y., 1978.

Jorgensen, Erik, "Urban Forestry in Canada," in *Proceedings of the 46th International Shade Tree Conference*, 43a–51a, 1970.

Millen, W., and R. Randall, "Urban Forestry—What Is It and Is There a Role for State and Private Forestry?", in-service document, USDA Forest Service, Policy Analysis, November 30, 1981.

Rooney, B., "Urban Forestry Steps Out," *American Forests,* 89(1):20, January 1983.

Society of American Foresters, "Directory of Urban Foresters, SAF Urban Forestry Working Group Publ., 7p., 1974.

Winters, Robert K., *The Forest and Man,* Vantage Press, New York, 393 p., 1974.

Wysong, Noel B., "Urban Forestry," *Arborist's News,* 37(7):76–80, 1972.

Zube, Ervin H., "The Natural History of Urban Trees," in *The Metro Forest, A Natural History Special Supplement,* 82(9), 1973.

2
DISTRIBUTION AND OWNERSHIP OF THE URBAN FOREST

The urban forest includes all woody vegetation within the environs of all populated places, from the tiniest villages to the largest cities. In this sense it includes not only trees within city limits but trees on associated lands that contribute to the environment of populated places—for example, greenbelts, municipal watersheds, recreation sites, and roadsides. Andresen has described the urban forests of megapolitan New York as "forests that invade the abandoned fields of the countryside and recolonize deserted lots and back alleys; forests that yield but do not surrender to the building contractor and the tract developer and the forests that, despite constant abuse, provide continual bounty." Thus, in New York State, the urban forest enlarges from rows of street trees and clusters of park trees in New York City to ever-broadening greenbelts in contiguous suburbs, to coalesce in the forests of the Catskills, Adironacks, and Allegheny Highlands (Andresen, 1975).

Nationwide, the urban forest covers an estimated 69 million acres (28 million ha).* As with New York, any heavily populated area in the nation has elements of the urban forest. However, a small town on the Great Plains—an oasis of trees surrounded by open prairie—is also an urban forest (Figure 2.1).

The distribution of urban forests in a particular geographic area is not difficult to determine. One has only to look at a map showing population centers to correctly determine that the larger part of the urban forest is in and near the larger cities. The distribution of the forest of individual cities, however, is much more complex and deserves a closer look.

Within city environs, it is necessary to look at the various land-use areas on which trees are likely to occur. The land-use data for metropolitan Dade County, Florida, shown in Table 2.1, are an example (Goodman, 1968). Nearly 35 percent of the total land area is in residential use, with most of this used for single family residences. Since residential areas are located either in natural forests or are generously planted to trees, it is reasonable to conclude that they constitute a major portion of the urban forest. Other land-use areas on which the urban forest most logically occurs are park and recreation areas (3.8 percent), transportation areas—particularly streets (24.6 percent), agricultural areas (2.2 percent), institutional lands (3.1 percent), and undeveloped areas (23.5 percent). Definition is made difficult by commercial and industrial areas, for instance, where trees are often less

*Precise figures for the urban forest area are not available, due in great part to the difficulty of defining its boundaries. In the first edition of this text, we recorded our estimate of 69 million acres (28 million ha). This was based on our judgment that approximately three percent ot the nation's land area could be considered urban forest and was derived from a summary of special land use areas in: Economic Research Service, USDA, *Major Uses of Land in the United States.* (Agricultural Economic Report No. 247) 1969, p. 26. We have seen no subsequent data to suggest the need for a significant alteration of this figure.

Figure 2.1 The urban forest encompasses all of the woody vegetation within the environs of all populated places. A small town on the Great Plains is also an urban forest.

abundant. There are occasional trees in these areas, however, that make a contribution to the urban environment.

In measuring the characteristics of Dayton, Ohio's vegetation configuration, Sanders and Stevens (1982) found that "58 percent of the city's land is not covered with artificial surfaces, and that 37 percent (over 6,000 acres) of the nonsurfaced land is covered with tree crowns. The city has almost three quarters of a square mile of exposed soil. Only single and two-family residential land use has more acreage with tree cover than grass cover. Wide variations in land cover characteristics exist throughout the city and across its land uses."

Any definition of the urban forest should be abstract as well as physical and should deal with all woody vegetation within cities and their environs. Thus, the data discussed above show only where one is most likely to find the urban forest. However, it can be everywhere within and around the cities.

For those who cannot perceive the urban forest as a partial abstraction (and must have acreages), a solid case can be made for including the total acreage of city environs. The rationale here is that roads, streams, lakes, meadows, and other open areas are not subtracted from acreages in National Forests or other large forested areas. By this reasoning we arrive at an

15

TABLE 2.1

Existing Land Use,
Metropolitan Dade County

Category	Acreage	Percent of Urban Area
Residential	(0.405 ha)	
Single family	39,526	31.0
Two family	1,800	1.4
Multifamily	1,783	1.4
Rooms	78	0.1
Camps	24	0.0
Trailers	328	0.3
Mixed[a]	708	0.6
Total	44,248	34.8
Commercial		
Retail	3,940	3.1
Mixed[a]	458	0.4
Total	4,398	3.5
Tourist		
Hotels	549	0.4
Motels	264	0.2
Mixed[a]	54	0.0
Total	870	0.6
Industry		
Extraction	339	0.3
Light manufacturing	639	0.5
Heavy manufacturing	434	0.3
Light storage	569	0.4
Heavy storage	250	0.2
Mixed[a]	343	0.3
Total	2,575	2.0
Institutional		
Education	1,909	1.5
Cultural	199	0.2
Medical	232	0.2
Religious	498	0.4
Public administration	641	0.5
Penal	6	0.0
Mixed[a]	351	0.3
Total	3,835	3.1
Parks and Recreation		
Parks	2,629	2.0
Play grounds	218	0.2
Golf courses	1,604	1.3
Cemeteries	344	0.3
Total	4,796	3.8

TABLE 2.1 *(continued)*

Category	Acreage	Percent of Urban Area
Transportation		
Terminals	6,784	5.3
Railroads	830	0.6
Utility	529	0.4
Streets	22,966	18.0
Parking	249	0.2
Mixed[a]	158	0.1
Total	31,516	24.6
Agriculture		
Groves	687	0.5
Crops	2,102	1.7
Mixed[a]	48	0.0
Total	2,834	2.2
Undeveloped		
Vacant	29,815	23.4
Glades	20	0.0
Marshes	78	0.1
Total	29,913	23.5
Water		
Lakes	1,104	0.9
Courses	1,290	1.0
Bays		0.0
Total	2,394	1.9
Total, all categories	127,381	100.0

[a]Indicates two or more categories of use on the same parcel of land.

Source: Metropolitan Dade County Planning Advisory Board and the Metropolitan Dade County Planning Department, *Preliminary Land Use Plan and Policies for Development,* The Department, Miami, p. 18, 1961.

estimated 69 million acres (28 million ha) as the total area of the urban forests in the nation.

Perhaps the most straightforward way of looking at the distribution of the urban forest is according to public and private ownership and public or private responsibility. We estimate that 30 percent of the trees in urban areas are publicly owned. The remainder is in private ownership.

Public Lands

Parks

In the urban setting, parks are probably the most relatable as forests (Figure 2.2). Indeed, many parks have been created from natural wooded areas,

17

Figure 2.2 Parks with trees and open spaces are probably the areas in the urban setting most easily identifiable as forests.

while others are magnificent examples of manmade forests. Notable among manmade forested parks is Goldengate Park in San Francisco. Planted on unstable sandy soil in the late nineteenth century, the forest of this park is now under intensive uneven age management—a vivid example of the application of conventional forestry to the urban environment.

Parks vary from tiny green spots in central business districts to large acreages, often beyond city limits. There are few parks without trees, even those dedicated to active recreation. Passive recreational areas are usually heavily forested. Most parks are publicly owned, but there are also recreation and other open areas on private lands owned by churches, private schools, industries, labor unions, and other organizations.

Most public parks are municipally owned and managed. However, there are occasional examples of county, state, and federally owned parks in or adjacent to cities. An excellent example of federally owned parkland is in and near Washington, D.C. where the Capital Mall, George Washington and Baltimore parkways, and other forested lands are administered and managed by the USDI National Park Service. These areas represent both planted and natural forests—the latter involving pristine stands of eastern hardwoods, wild in every sense, literally surrounded by millions of people.

18

Figure 2.3 Street rights-of-way constitute a substantial part of public urban forest lands.

Street rights-of-way

Street rights-of-way constitute a substantial part of public urban forest lands (Figure 2.3). These are strips adjacent to streets or medians between divided boulevards. They are often called treelawns, parkways, or parking strips. Streetside rights-of-way vary in width and often provide space for sidewalks. There is usually space for a single row of trees except that median areas are often wide enough for trees, shrubs, and other landscape design features. Streetside tree spacing varies according to lot width, driveways, utilities, and other spatial factors; but 50 ft (15¼m) apart or 200 trees per street mile (1.61 km) is generally the recommended standard in residential areas. It must be stressed that the streetside tree situation varies greatly within most cities. The factors responsible for these variations are discussed in Chapter 6.

Highway and railroad rights-of-way

There are virtually no cities or towns in modern America that are not served by public highways or railroads. Most cities have one or more railroads and perhaps several county, state, and federal highways. The land adjacent to

Figure 2.4 Lands adjacent to highways support a significant part of the urban forest.

Figure 2.5 Grounds next to public buildings are important parts of the urban forest.

20

these arteries, at intersections, and in medians often supports a significant part of the urban forest (Figure 2.4). Along some highways, particularly those maintained by counties or states, tree situations are similar to those along streets. However, federal highways (often called interstates, freeways, or trafficways) generally involve large land areas where trees, shrubs, and other landscape plants have been planted. These areas, particularly intersections, have often been the focus of major urban forestry projects. Intersections, or "cloverleafs," frequently occupy large areas with residual open land, often with site conditions conducive to excellent tree growth. In many cases, especially in the South, these areas have either been planted to trees, or existing forest species managed for interim benefits and ultimate harvest for forest products. "Plant digs" have been organized in forested areas to be cleared for freeway construction. Individuals may dig trees for replanting, or may harvest trees for fuelwood or other "household" purposes. In return, they may be asked to agree to help replant seedlings in open areas after construction.

Railroad rights-of-way make only a minor contribution to the urban forest environment as they were rarely landscaped in the past. In recent years, however, abandoned roadbeds and yard areas have been converted to bicycle trails, parks, and other recreation areas with appropriate landscape plantings.

Public buildings and grounds

Grounds adjacent to public buildings (e.g., schools and colleges, hospitals, auditoriums, museums, penal institutions, courthouses, and utility plants), are important parts of the urban forest (Figure 2.5). Quite often these grounds are well landscaped with generous plantings of trees and shrubs. Other public grounds may be used as military installations, cemeteries, airports, golf courses, and nurseries (Figure 2.6). These segments of the urban forest may be owned and managed by all levels of government and represented by a myriad of departments and agencies.

Extraterritorial lands

The urban forest often goes beyond the populated limits of the city. Greenbelts, groves, forest preserves, and even vegetated landfills, although often removed from the city proper, must be considered as part of the urban forest. These areas are usually publicly owned and often dedicated to a primary use with secondary multiple uses. These forests are extremely valuable to the urban environment as they may provide watershed protection, recreation, scenery, solace, or a place for disposal of waste products.

Although only partially extraterritorial, the Chicago area forest preserves are an excellent example of such lands. The Cook County Forest

Figure 2.6 Cemeteries are often forests.

Preserves were authorized by the Illinois State Legislature in 1913 and organized as a Forest Preserve District in 1915. The Preserves total more than 47,000 acres (19,000 ha) in many tracts of various sizes located north, west, and south of Chicago. They comprise a network of forested lands dedicated as naturalistic sanctuaries for the education, pleasure, and recreation of the public (Otte and Vlasin,1963).

Most tracts are maintained in primitive state. Their interiors are accessible only by trails established for walking, bicycling, or horseback riding. Although they are not intended as parks in the traditional sense, there has been intensive development of recreation facilities on exterior areas.

Many of the preserves are within the corporate limits of Chicago and its suburban towns. The many tracts are connected by public roads or trails on strips of public land.

The Forest Preserve District is governed by a board of 15 Forest Preserve Commissioners. Over many years the Board of Commissioners has withstood near-relentless pressure by individuals, groups, and public agencies seeking forest preserve lands for conventional parks, schools, cemeteries, housing developments, and commercial purposes.

An occasional village forest still remains, particularly in the eastern United States. Originally set aside as commons to provide fuel and other wood products for village residents, these forests are now most often parks or other recreation areas.

Recently solid waste disposal areas, when filled, have been planted to trees and converted to a variety of recreational uses. Such reclaimed landfills are reasonably certain to add substantially to the urban forest in the future.

In addition, there are in most urban areas tracts of woodlands left as islands in residential or other areas because of zoning restrictions relating to soil stability, flooding, or other restricting factors. Many of these lands are left in limbo as to management responsibility. They provide wildlife habitat and "woods experiences" for people, particularly children, who live adjacent to them. In most cases, these areas could provide more public benefits, even forest products, if administered and managed differently.

Riparian areas

Riverfronts, canals, channel diversions, marshes, lakeshores, and even seashores are parts of the urban forest. (Figure 2.7). Quite often these lands are developed as parks or recreation areas. In most cases, they cannot be

Figure 2.7 Riparian areas such as this riverfront recreation site contribute to the urban forest.

Figure 2.8 The largest area of private ownership in the urban forest is in residential districts.

developed for residential or commercial uses because of flooding or other site limitations. They do serve, though, as greenbelts and open spaces within cities. Such water-related areas are extremely varied, and trees are important elements when their presence is not inconsistent with the primary use of the land. A magnificent example of a riparian urban forest is along the Potomac River and Rock Creek in Washington, D.C. These areas are parklands and were considered in the previous discussion of parks. They are, however, water-related and vary form "primitive" forests along the river and the old C and O Canal to intensively used areas around the tidal basin.

Private Lands

The private portion of the urban forest occurs on all kinds of residential, commercial, and industrial lands. The largest area is in residential districts either in natural forested areas or where property owners have often created magnificent forests (Figure 2.8). Residential areas vary from highrise apartment complexes to single family dwellings, and trees are associated with

24

both settings except in infrequent situations where site factors are too limiting. Trees are a part of the appeal of home ownership, with well-treed lots adding substantially to property values. In open areas, a sturdy shade tree is one of the first introductions to a new house. As suggested by Dade County, Florida data, residential areas constitute a substantial part of urban land use. Although precise figures for the total urban forest area do not exist, it appears reasonable that 20 to 30 percent of the nation's urban forests are in residential areas.

Commercial areas make up a less significant part of the urban forest. This is because of the relatively small land area devoted to commercial purposes and relative lack of trees there. Tree plantings and other landscaping vary greatly but appear to have increased during the last two decades, although perhaps slowed in more recent years because of general economic conditions.

Industrial districts are often more utilitarian than beautiful. However, many industries contribute substantially to the total urban forest with landscaped grounds, screen plantings, and outdoor park and recreation areas for employees. Many cities, seeking to attract industry, have developed industrial parks, or areas with services and utilities, where companies can locate. Tree plantings are sometimes made in these areas as an esthetic inducement to industry and are often required of industry after construction.

There has been a trend in recent years for corporate headquarters and other business offices to locate in suburban areas. Developers of such facilities appeal to their customer's sense of sylvan values with names such as "Corporate Woods," "Professional Grove," and "Technology Forest." There is a general belief that such settings contribute to employee productivity.

Public or Private Management Responsibility

Ownership of land on which the urban forest grows is never total and absolute. Some rights must be given for the benefit of society. Private property rights are commonly given (taken) in the form of easements for streets, power lines, sewers, and other utilities. Such easements often restrict the planting of trees to certain areas and numbers and require pruning and other management practices to limit interference. Conversely the public may give rights and/or responsibilities to private property owners. Title to trees growing on street rights-of-way sufficient to claim injury damage is commonly given to adjacent private property owners. The responsibility for pruning and care of streetside trees is also often transferred to property owners. The transfer of rights and responsibilities is generally prescribed

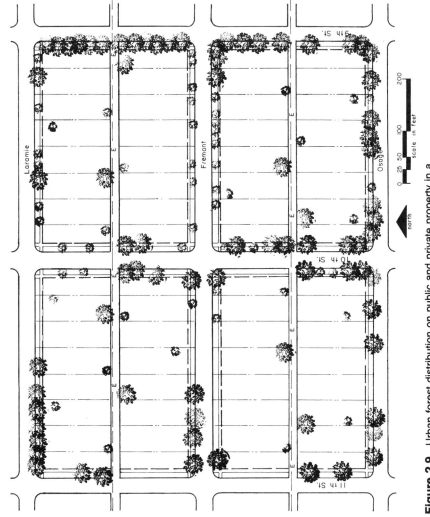

Figure 2.9 Urban forest distribution on public and private property in a selected area of Manhattan, Kansas.

26

by ordinance or code. In many cities and towns, however, streetside tree responsibility is based on policy or tradition rather than ordinance.

These factors are illustrated by a four-block section of Manhattan, Kansas (Figure 2.9). This area was platted prior to 1900 and is typical of older residential neighborhoods in many cities. The blocks are divided into eight 50 x 150-ft (15.25 x 47.75m) lots on either side of a 15-ft (4.57-m) alley. The streets are 30 ft (9.14m) wide with 15-ft (4.57-m) parkways on both sides. Sidewalks, 4 ft (1.22m) in width, are located near the private property side of the parkways, leaving 9-ft treelawns.

Treelawns are part of street rights-of-way and are owned by the city. Title to trees on them is held by adjacent property owners. City ordinances give the city the right to plant, prune, spray, and otherwise manage parkway trees, but also allows property owners to do so by permit. Dead trees are removed by the city and new trees are planted where appropriate. However, watering and other postplanting care are by policy the responsibility of property owners.

Overhead power lines run along the alleys on city property. Pruning of interfering trees on adjacent private property is the right of the city. Responsibility for pruning, however, is transferred to the utility company. The right of owners to manage trees on their property is further influenced by city ordinances that require removal at the owner's expense of dead or diseased trees.

Ownership and distribution are of particular significance to the management of the urban forest. Even in a small city or town the urban forest might be owned by several thousand private property owners plus numerous departments of the city, county, state, and federal governments. In addition, there are transferred responsibilities by easements, ordinances, and policies. Such a diversity of ownership and responsibility both facilitates and complicates management of the urban forest. It offers opportunities for intensive care by individuals of small segments of the forest but poses definite problems when management systems that apply to the total forest are considered. Management of the urban forest is discussed in detail in Chapter 6.

BIBLIOGRAPHY

Andresen, John W., "New York's Urban Forests," NAHO, The University of the State of New York, The State Education Department, Vol.8, No. 1, p. 6, Spring 1975.

Goodman, William I., *Principles and Practice of Urban Planning,* International City Managers' Association, Washington, D.C., p. 122, 1968.

Otte, Robert C. and Raymond D. Vlasin, "Districts That Manage Resources," *A Place to Live,* The Yearbook of Agriculture 1963, U.S. Government Printing Office, Washington, D.C., p. 434, 1963.

Sanders, R.A., and J.C. Stevens, "The Urban Forest of Dayton, Ohio: A Preliminary Assessment," In-Service Report, USDA Forest Service and SUNY College of Environmental Science and Forestry, Syracuse, N.Y., 1982.

3
COMPOSITION OF
THE URBAN FOREST

The urban forest can be both natural and manmade. In total area, it is perhaps the largest of all planted forests in the nation. Precise figures are not available, but we estimate that approximately 45 million acres (18.225 million ha) have been planted in urban and recreation areas, at public installations and facilities, and along roads and other transportation areas. The urban forest includes also a large area of native woodlands both within and adjacent to cities and towns. In many cases, native forests have been urbanized as cities have grown into them (Figure 3.1). Often in these situations, the original forests have been so altered by the supplemental planting of introduced species that they have taken on the characteristics of manmade forests. In areas such as the Great Plains or desert Southwest, the urban forest may be entirely manmade.

It is virtually impossible to discuss the total composition of the urban forest. It is more meaningful to discuss specific areas of the nation. The urban forests of New England, for example, are composed of generally similar species that are quite different from those in the urban forests of the Gulf States, Lake States, Great Plains, Rocky Mountains, or West Coast. However, even within these areas considerable climatic differences occur that influence the composition of the forests.

A common approach to urban forest composition is the use of plant hardiness zones. These zones were delineated by the Arnold Arboretum and are based on annual average minimum temperature. Species of trees and other plants generally adaptable to the climatic conditions of each zone have been identified. A map showing hardiness zones is shown in Figure 3.2*. Hardiness zones have the obvious limitation, however, of not taking into account more localized site factors such as soils and exposure. These factors are discussed in Chapter 5.

In forested regions of the United States, urban forests are composed largely of indigenous species. This is because cities have been built inside forests, and native species are often those generally most available for planting. Exceptions are in regions of sparse natural forests where introduced trees must be used, or in subtropical areas where exotic species are so easily adaptable. In forested regions, however, there are more exotic species within cities, which give way to indigenous species in suburban and exurban areas.

The composition of urban forests is influenced primarily by physical elements. The limiting factor is which species will grow in an area and ranges from hundreds of species in the south to a handful in the high elevations of the Rocky Mountains. Moreover, within the limits imposed by nature are the personal, social and economic factors that influence human

*For a detailed listing of tree species by zones, see Donald Wyman, *Trees for American Gardens*, MacMillan, New York, 1965.

Figure 3.1 Native forests have become urbanized as cities have grown into them.

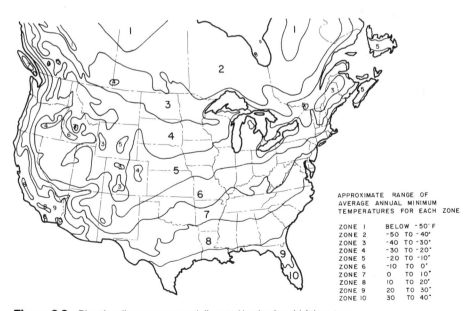

APPROXIMATE RANGE OF
AVERAGE ANNUAL MINIMUM
TEMPERATURES FOR EACH ZONE

ZONE		
ZONE 1	BELOW	-50° F
ZONE 2	-50 TO	-40°
ZONE 3	-40 TO	-30°
ZONE 4	-30 TO	-20°
ZONE 5	-20 TO	-10°
ZONE 6	-10 TO	0°
ZONE 7	0 TO	10°
ZONE 8	10 TO	20°
ZONE 9	20 TO	30°
ZONE 10	30 TO	40°

Figure 3.2 Plant hardiness zones as delineated by the Arnold Arboretum.

choice. While often interrelated, these factors may be summarized as purpose or function, popular species, public control, socioeconomic factors, mobility, and nostalgia.

Purpose or Function

Composition of the urban forest is influenced largely by the purposes its various segments are intended to serve—such as shade, visual screening, wind protection, esthetics, or production of wood products. For example, a picnic area in a public park would probably have large deciduous shade trees. Plantings for visual screening or wind protection would be composed mainly of evergreens. Streetside plantings for shade or esthetics are often colorful deciduous trees. A single tree might be chosen for size, form, color, or another characteristic to fill a particular need in the landscape. Large stands of naturally occurring trees might serve as esthetic or privacy elements on large estates, or might actually be producing forests at the urban fringe. Planted forests within urban areas might also be for Christmas trees, fuelwood, or other products.

Popular Species

The composition of urban forests often reflects the popularity of certain species. In the East and Midwest, remaining American elms attest to the species' popularity prior to the spread of Dutch elm disease. Sugar maples are popular in the Northeast; mimosas in the South; and Modesto ashes on the West Coast. Other currently popular species over a wide geographic range are honeylocusts, pin oaks, sweetgums, and red maples. A species popularity is often based on practical reasons such as its form and strength, as in the case of American elms. It might also be based on local availability. And local availability is influenced strongly by economics of nursery production—easily propagated species with lower nursery maintenance requirements are most available.

More often, however, species popularity is based on brilliant leaf or blossom color or other desired characteristic. Maples, oaks, gums, and crabapples are often planted for these reasons. Popularity is often created by advertising and promotion by nurseries and landscape firms and is related to economics as stated above. There are often, for example, full page ads for maples, elms, poplars and other species in popular magazines. Popularity is also influenced by the recommendations of forestry and horticulture specialists, lawn and garden editors, and others. Unfortunately, a species' popularity often leads to planting mistakes, because a tree chosen only for leaf color or one other characteristic may be unsuitable for the planting site. A common example of this occurred in the Midwest where blue spruces were planted on front lawns in near total disregard for other landscape considerations. Another common example is the placement of sugar maples (red leaves) on hot, dry, exposed sites.

Public Control

Planting programs instituted by city forestry departments have the most controlling influence on the composition of urban forests. Directed principally to street rights-of-way, parks, and other public grounds, these programs range from recommendations or regulations intended for adjacent property owners to total implementation by city crews. Many cities have "official" street lists of trees from which adjacent property owners may choose. Other cities prohibit by ordinance the planting of certain "undesirable" species. Streetside planting may also be controlled or influenced by subdivision regulations that require developers or property owners to plant trees. The most positive control, however, is in cities where all streetside tree planting is done by municipal forestry departments.

Less formal and most applicable in small cities that do not have forestry departments are planting projects sponsored by civic and service clubs, garden clubs, youth groups, and other organizations. Such projects are usually intended for parks, school grounds, highway entrances, or other public areas.

Composition is also influenced by various legal factors concerning urban development in natural wooded areas, ranging from the legal basis for urban planning, to land use zoning, to site planning. These factors will be considered in more detail in Chapter 8.

Socioeconomic Factors

The composition of urban forests varies greatly according to economic areas within particular cities. In areas of low resident income, the urban forest is often composed of declining older trees remaining from times of greater neighborhood prosperity. These areas are generally closer to the inner city, but were once at the "wooded fringes" where the more affluent lived. That trees and wooded lots have long been associated with upper and middle class residential values was well stated by Zube (1973):

> *During the time of the Roman Empire, the Renaissance, or even later nineteenth-century America, those who could afford it lived on the fringe of the city absorbed by nature—although sometimes the trees were made to conform to man's idea of a more appropriate form by pollarding or topiary work. Go to almost any American metropolis today and seek out the most prestigious residential areas—you will probably find that most are in wooded sections. Names also give an indication: Forest Hills, Woodland, Lake Forest, and a host of others that explicitly or implicitly equate trees with a quality living environment.*

In these areas of middle and upper income the urban forest is generally well planted and well tended, reflecting the options of affluence. It is also in areas of higher income that most natural woodlands on private property are found. Large lots and even large acreages are common—premium sites for home construction. Schmid (1975) observed that in such areas of Chicago, homeowners must be interested in or at least tolerant of native vegetation: "At the very least, a resident must feel no obligation to convert the whole of his property into lawns and flower beds in order to acquire neighborhood respectability."

As indicated, the growth of American cities has often caused these areas to be converted to uses other than residential, or "decline" into districts of lower income. Tree planting and care simply do not command high priority in areas of low income—either by residents or local government. When new trees are planted, there is a tendency to select fast-growing species such as silver maples, willows, and poplars. Tree-of-heaven is also sometimes planted, but more commonly establishes itself naturally on vacant lots and other untended areas.

Mobility

Although difficult to separate from socioeconomic factors, the mobility of Americans also has an influence on composition of urban forests. People in low and middle income areas tend generally to own homes for only a few years. Tenure is longer in districts of higher income. People who view their residence in a particular location as temporary tend not to plant trees. However, the frequent resale of property increases the chances of ownership by individuals who will plant trees, often in the hope that such plantings will increase property values. When trees are planted, fast-growing species are most often used because there is not time to wait for oaks, hickories, and other slower growers.

Nostalgia

Nostalgia for the familiar has had a profound influence on the composition of urban forests. Trees from the "old country" or "back East" were often brought by settlers. Although many species could not tolerate the new sites to which they were introduced, many of them flourished. Often such introductions were for utilitarian as well as esthetic purposes as attested by the early introductions of apple, pear, peach, and nut trees.

Occasionally a love of trees once known became a crusade, as in the case of J. Sterling Morton, founder of Arbor Day. Remembering fondly the large trees of his native New York State, Morton set about to have the planting of trees on the Nebraska plains made an official act. His zeal not only changed the Nebraska landscape but has influenced the urban forests of virtually every city in North America.

Other Factors

Composition is also influenced by forest insect or disease epidemics. Dutch elm disease is the classic example of recent times. In many urban areas, it has reduced American elms to an insignificant factor and has resulted in an urban forest totally different in size and composition. The history of Dutch elm disease is replete with cases of residential streets once bordered and canopied with graceful elms now exposed and stump-cluttered. Chestnut blight impacted the urban forests of the East—although to a lesser degree—in earlier years. Although it is obviously too soon to assess the total effects of gypsy moth epidemics, changes in forest composition are apparent as vulnerable species succumb to sometimes repeated defoliation.

As indicated in the previous chapter, land uses within urban areas influence not only distribution but composition of urban forests. This was illustrated by a streetside tree inventory of Syracuse, New York (Richards and Stevens, 1979). They found an average of 10.7 trees per 1,000 feet of streetside planting strip, but stated that interpretations of average densities for neighborhoods requires consideration of neighborhood characteristics:

The very low street tree density in the Lake Industrial area reflects its nonresidential character. The low figure for the Central Busi-

35

ness District, nearly all young trees, reflects the difficult and specialized conditions for street trees there. Similarly, the rather low density for the Westside is partly explained by its commercial component. On the other hand, residential areas in Thurber and Far West are better stocked than the average figures suggest because of significant undeveloped areas included there.

Their further comments relate urban forest composition not only to land use but to other factors previously discussed:

Densities of trees over 16 inches in diameter, most of the older tree population, suggest the relative importance of the elm losses among neighborhoods. Conversely, densities of trees under 8 inches indicate the relative intensity of recent replacement efforts. Older residential neighborhoods (largely pre-1908 development) that were not planted heavily to elms, now have relatively high densities of street trees over 24 inches. Younger neighborhoods generally have above-average densities in the 16–24-inch class.

The composition of the urban forest is also, of course, influenced by other phenomena such as hurricanes, tornadoes, floods, and fires. It is perhaps trite, but true, that the urban forest reacts to all of the whims of nature and man.

Manhattan, Kansas

All of the factors just discussed have influenced the composition of the urban forest of Manhattan, Kansas. Manhattan is located within hardiness zone 4. It has a typical continental climate with warm summers and moderately cold winters. Normal low temperatures are 30°F (-1°C) in January and normal highs are 82°F (27°C) in July. Extremes have ranged from -30°F (-34°C) to $+112$°F(44°C). Average rainfall is 32 in. (81.28 cm) with approximately 70 percent of it falling during the growing season. In summer, the rate of evapo-transpiration is moderately high. Prevailing summer winds are from the south. Periods of high winds can be expected particularly in March, April, and May.

Manhattan is located in the Flint Hills region of Kansas. The original city was built in the bottomlands of the Kansas River. Subsequent development of the city took place on adjacent uplands. Lower slope and bottomland soils are moderately deep silty clays on undisturbed sites. Upland soils are shallow, stony, and gravelly. All soils tend to be alkaline.

Manhattan has a population of 29,500. It was settled in 1855 by farmers and traders of varying ethnic backgrounds. The city developed primarily around the Bluemont College, which later became Kansas State College (now Kansas State University).

The urban forest of Manhattan is mostly manmade, although some development of the city has extended into naturally wooded ravines and hillsides. The urban forest extends beyond the city limits to include public recreation areas adjacent to a large lake owned by the federal government. The forest within the city limits can be divided into three general areas: (1) older; (2) intermediate; and (3) new (Figure 3.3). The "older" area includes the originally platted city plus its expansion on the high river bottoms east and south of the university. The "intermediate" area is west and north of the campus and was developed generally from 1940 to 1960. The "new" area

Figure 3.3 The urban forest of Manhattan, Kansas can be divided into three areas: (1) older; (2) intermediate; and (3) new.

TABLE 3.1

City Street Trees
City of Manhattan, Kansas
Older Area
July 1, 1975

Species	Number of Trees	Average Age	Average Diameter (in.) (2.54 cm)	Percent of Species Total				Percent of Total Trees
				Good(1)	Fair(2)	Poor(3)	D & D(4)	
American elm	2158	60	19	31	25	24	20	33
Hackberry	1090	60	16	59	31	10	0	18
Siberian elm	376	40	14	4	33	59	4	6
Hard maple (sp.)	328	10	4	74	5	18	3	5
Green ash	320	15	6	48	31	21	0	5
Pin oak	320	40	15	69	10	20	1	5
Sycamore–planetree	302	20	8	76	9	15	0	5
Honeylocust	288	20	9	71	8	20	2	4
Silver maple	280	15	10	39	13	48	0	4
Red oak	218	10	5	56	23	13	8	3
Redbud	138	10	4	78	2	20	0	2
Black walnut	130	55	15	34	54	12	0	2
Hybrid elm	82	15	6	7	56	37	0	1
aMiscellaneous	466							7
Totals	6496							

(1) *Good:* Healthy, vigorous tree. No apparent signs of insect, disease, or mechanical injury. Little or no corrective work required. Form representative of species.

(2) *Fair:* Average condition and vigor for area. May be in need of some corrective pruning or repair. May lack desirable form characteristic of species. May show minor insect injury, disease, or physiological problem.

(3) *Poor:* General state of decline. May show severe mechanical, insect, or disease damage, but death not imminent. May require repair or renovation.

(4) *Dead or dying:* Dead, or death imminent from Dutch elm disease or other causes.

---

[a] Less than one percent each: bur oak, tree of heaven, flowering crab, sweetgum, basswood, black locust, Kentucky coffeetree, eastern redcedar, chinquapin oak, hawthorn, mimosa, catalpa, pine (sp.), cottonwood, willow, Osage-orange, purpleleaf plum, white birch, mulberry, fruit (sp.), Japanese pagodatree, shingle oak, English oak, Russian olive, ginkgo, boxelder.

is located on the uplands to the west. In July 1975, a complete inventory was made of streetside trees in each area.

Older area

The "older" area of Manhattan supports 6496 street trees composed of more than 40 species (Table 3.1). In spite of sustained losses from Dutch elm disease, American elm is still the most abundant species, comprising 33 percent of the total population. (A 1948 inventory showed that 66 percent of them were American elm.) Hackberry is the second most common species with 18 percent. The remaining tree population is well distributed, with no single species making up more than six percent of the total. The American elms and hackberrys have an estimated average age of 60 years. All of the "original" street trees were planted by adjacent property owners as there were no city-sponsored planting efforts until the late 1940s. The choice of species was based largely on the local availability of elms and hackberrys. Both species are native to the area, and wild seedlings and saplings were frequently transplanted. A local commercial nursery also promoted these species. Because there was no forewarning of epidemic disease, the selection of elms and hackberrys could hardly be faulted. Because of their hardiness and vase-shaped ascending form, both species were admirably suited for streetside planting.

There is little clear evidence that nostalgia was an important factor in street planting except for a few large black walnuts that were planted along a farm lane that later became a city street. These were planted by an individual who remembered fondly the large walnut trees in the eastern states and who also relished black walnut nutmeats. Ethnic factors seem not to have been important since significant numbers of a particular species that could be associated with "mother" countries do not exist.

The "older" area of Manhattan includes sections of both high and low resident income, and the streetside tree data are not separate. Thus, conclusions can be based on observations only. The lower income areas appear to have fewer trees. Elm losses have been high but significant replacements have not been made. Streetside planting is done by city crews while postplanting care is the responsibility of adjacent property owners. Because of generally poor care exercised by residents, the city forestry division has discontinued planting unless it has a prior agreement with the property owners for maintenance. Tree planting request and care agreement forms are published in the local newpaper prior to the planting season. Few forms are submitted by residents of low income sections of the city.

Forty species of streetside trees is a relatively high number for Kansas towns. (The Department of Forestry at KSU has inventoried 110 towns.) Observations suggest that there is also a generally greater number of additional species on private property than in other towns. This can be reasonably attributed to the proximity to the Departments of Horticulture and Forestry at Kansas State University. The experimental and extension work of staff horticulturists and foresters has had both a direct and indirect influence on the composition of Manhattan's urban forest. The department

TABLE 3.2

Recent Street Tree Planting in the
Older Area of Manhattan
(Trees Less Than 3 In. Diameter)

Species	Number of Trees	Percent of Total
Hard maple (sp.)	105	24
Green ash	50	12
Red oak	52	12
Sycamore–planetree	52	12
Silver maple	43	10
Redbud	19	4
Sweetgum	18	4
Pin oak	14	3
Hackberry	8	2
Honeylocust	7	2
Basswood	7	2
Flowering crab	8	2
Fruit (sp.)	8	2
Siberian elm	4	1
Black walnut	5	1
Hybrid elm	5	1
Cottonwood	4	1
American elm	8	>1
Eastern redcedar	1	1
Catalpa	1	1
Bur oak	3	1
Willow	1	1
Mulberry	1	1
Russian-olive	2	1
Ginkgo	1	1
Black locust	1	1
Mimosa	3	1
Purpleleaf plum	1	1
White birch	1	1
Japanese pagoda-tree	1	1
	429	

TABLE 3.3

City Street Trees
City of Manhattan, Kansas
Intermediate Area
July 1, 1975

Species	Number of Trees	Average Age	Average Diameter (in.) (2.54 cm)	Percent of Species Total				Percent of Total Trees
				Good(1)	Fair(2)	Poor(3)	D & D(4)	
Honeylocust	648	20	10	80	18	2	0	20
Sycamore–planetree	504	20	11	90	6	4	0	16
Green ash	322	20	10	61	30	8	0	10
Hackberry	250	25	12	69	22	8	1	8
American elm	164	30	16	45	47	5	4	5
Red oak	158	20	9	90	6	4	0	5
Siberian elm	158	20	11	28	34	38	0	5
Flowering crab	128	15	6	91	9	0	0	4
Silver maple	110	15	8	58	35	7	0	3
Hard maple (sp.)	106	10	4	77	17	6	0	3
Hybrid elm	90	20	10	36	49	15	0	3
Redbud	66	10	4	73	24	3	0	2
Pine (sp.)	62	20	9	81	16	3	0	2
Goldenraintree	50	20	8	88	12	0	0	2
Pin oak	50	20	10	88	0	12	0	2
Sweetgum	50	10	4	56	36	8	0	2
Japanese pagoda-tree	46	20	10	48	43	9	0	2
Eastern redcedar	40	20	10	15	55	30	0	1
ªMiscellaneous	166							5
Totals	3194							

(1) *Good:* Healthy, vigorous tree. No apparent signs of insect, disease, or mechanical injury. Little or no corrective work required. Form representative of species.

(2) *Fair:* Average condition and vigor for area. May be in need of some corrective pruning or repair. May lack desirable form characteristic of species. May show minor insect injury, disease, or physiological problem.

(3) *Poor:* General state of decline. May show severe mechanical, insect, or disease damage, but death not imminent. May require repair or renovation.

(4) *Dead or dying:* Dead, or death imminent from Dutch elm disease or other causes.

*Less than one percent each: white birch, bur oak, cottonwood, willow, basswood, baldcypress, black walnut, Kentucky coffeetree, mimosa, mulberry, black locust, fruit (sp.), paw paw, hawthorn, catalpa, Russian-olive, chinquapin oak, boxelder, Osage-orange, and white poplar.

TABLE 3.4

City Street Trees
City of Manhattan, Kansas
New Area
July 1, 1975

Species	Number of Trees	Average Age	Average Diameter (in.) (2.54 cm)	Percent of Species Total Good(1)	Fair(2)	Poor(3)	D & D(4)	Percent of Total Trees
Sycamore–planetree	380	10	6	42	40	18	0	15
Hard maple (sp.)	222	10	4	59	32	5	4	9
Honeylocust	220	15	7	36	47	17	0	9
Green ash	152	10	5	38	47	12	3	7
Fruit (sp.)	148	10	4	76	24	0	0	6
Hybrid elm	138	15	7	45	43	12	0	6
Silver maple	108	10	4	50	43	7	0	5
Pin oak	104	15	7	46	52	2	0	4
Red oak	104	10	6	75	23	2	0	4
Hackberry	92	10	5	24	44	30	2	4
Sweetgum	92	10	3	65	26	9	0	4
American elm	80	35	15	5	27	50	18	3
Pin (sp.)	70	10	4	97	0	3	0	3
Basswood	66	10	3	33	67	0	0	3
Bur oak	62	10	4	52	29	16	3	3
Mulberry	48	10	3	54	42	4	0	2
Eastern redcedar	32	20	9	38	31	31	0	1
Siberian elm	32	30	14	6	50	44	0	1
Redbud	30	10	3	73	7	13	7	1
White birch	24	10	3	100	0	0	0	1
aMiscellaneous	130							6
Totals	2334							

(1) *Good:* Healthy, vigorous tree. No apparent signs of insect, disease, or mechanical injury. Little or no corrective work required. Form representative of species.

(2) *Fair:* Average condition and vigor for area. May be in need of some corrective pruning or repair. May lack desirable form characteristic of species. May show minor insect injury, disease, or physiological problem.

(3) *Poor:* General state of decline. May show severe mechanical, insect, or disease damage, but death not imminent. May require repair or renovation.

(4) *Dead or dying:* Dead, or death imminent from Dutch elm disease or other causes.

aLess than one percent each: white birch, bur oak, cottonwood, willow, basswood, baldcypress, black walnut, Kentucky coffeetree, mimosa, mulberry, black locust, fruit (sp.), paw paw, hawthorn, catalpa, Russian-olive, chinquapin oak, boxelder, Osage-orange, and white poplar.

operated an experimental nursery for several years from which excess trees were made available to the public. Department personnel have served as consultants to the city forestry division on numerous occasions. Information on tree selection, planting, and care has long been readily available to Manhattan residents.

The planting program of the city forestry division does not preclude property owners from also planting streetside trees. A planting permit is required, but the system is poorly enforced, and much streetside planting is done in ignorance of, or deference to, the permit "requirement." Of 30 species (429 trees) planted in recent years, 24 percent are hard maples (Table 3.2). This is an apparent reflection of the popularity of hard maples, primarily because of their brilliant fall leaf color. Several magnificent specimens of these species on a prominent estate within the city have contributed greatly to this popularity.

Intermediate area

There are 3,194 streetside trees in the "intermediate" area of Manhattan (Table 3.3). There are 38 species present, most of which were planted during the past 30 years. The thornless honeylocust is the most common (20 percent), followed by the planetree (16 percent), green ash (10 percent), and hackberry (8 percent). No other single species makes up more than 5 percent of the total population. Most streetside trees were planted by the city as the area was developed. This program was undertaken by the city forestry division shortly after World War II. With counsel from the university, planting standards and policies were developed and species were selected according to general site conditions. It can be argued that there is a slight overabundance of honeylocusts and planetrees here, but thus far neither species appears to have serious problems.

Three percent of the streetside trees in this area are hybrid elms. These are Christine Buisman elms, *Ulmus carpinafolia* H.V., which were highly touted replacements for the American elms lost to the Dutch elm disease. They were planted by the city forestry division in response to the sentiment of the mid-1960s that elms should not disappear from the landscape. Although apparently highly resistant to Dutch elm disease, these particular trees appear to lack the qualities that made American elms such popular shade trees. They lack the ascending form of American elms and are subject to severe annual attacks by elm leaf beetles.

New area

Essentially the same factors that influenced the composition of streetside plantings in the intermediate area are present in the new area (Table 3.4). Most trees have been planted by city crews. Species composition is similar

46

with minor differences in percentages. The increased number of hard maples is perhaps noteworthy. Their presence appears to reflect the same popularity that is responsible for their recent planting in the older area of Manhattan. There are relatively few of them in the intermediate area only because there has not been the need for recent planting. Most of the hard maples here were planted by adjacent property owners. The city forestry division plants few maples in new areas because these species have difficulty withstanding the harsh elements of exposed sites.

BIBLIOGRAPHY

Richards, N.A., and J.C. Stevens, *Streetside Space and Street Trees in Syracuse—1978,* SUNY College of Environmental Science and Forestry, Syracuse, N.Y., p. 19, 1979.

Schmid, James A., *Urban Vegetation,* Research Paper No. 161, University of Chicago, Department of Geography, p. 63, 1975.

Zube, Ervin H., "The History of Urban Trees," *The Metro Forest,* A Natural History Special Supplement, p. 50, November 1973.

4
BENEFITS OF
THE URBAN FOREST

The urban forest is important to the city dweller in many ways. Its trees provide shade, beauty, and a long list of other benefits. In most instances these benefits are taken for granted. Indeed, the urban dweller may not even be aware of many or even relate to them. The various benefits can be grouped under the following four broad categories:

1. Climate amelioration
2. Engineering uses
3. Architectural uses
4. Esthetic uses

Climate Amelioration

The major elements of climate that affect us are solar radiation, air temperature, air movement, and humidity; and we have comfort zones associated with the interactions of these four elements. We can be too hot, too cold, or just right. This comfort zone of "just right" varies according to individual, sex, age, and the particular climate to which one is adjusted. From an engineering aspect, we control this comfort zone very precisely in buildings. With a twist of a knob or push of a button, we can regulate indoor temperature, light, humidity, and air movement. If only similar controls could be applied to the out-of-doors environment! To a degree they can be; by the proper use of trees and shrubs, a microclimate can be created that may ameliorate the climate sufficiently for us to be comfortable.

Temperature modification

Human confort essentially depends on the factors that affect skin temperature and the perception of heat and cold. The optimum core temperature for a human body is 98.6°F (37°C). Discomfort occurs when thermal energy is either lost or gained in relation to this optimum. This energy transfer, when carried to extremes, can result in either sunstroke or hypothermia. We generate heat through our metabolic processes. Inactive persons generate about 50 kcal of heat per hour. With physical exertion the amount of heat generated increases; to maintain comfort in either case, this heat must be dissipated.

Heat is dissipated from the body in three ways: radiation, convection, and evaporative cooling. Heat energy is radiated by people and their surroundings. This occurs through infrared thermal radiation from the skin or other surfaces. A two-way flux of radiation is involved. First, if the surroundings are cooler than the human body, the flux will be away from the body and cooling will result. Second, if the surrounding surfaces are warmer, the net flux of energy will be toward the human body. Heat is also transferred by convection. If the air is cooler than the skin or clothing, convective transport of heat will be toward the air. If the air is warmer, the transport will be from the air to the body. Wind also plays a very important role in convective heat transfer. It will speed up the transfer and can be especially critical where air temperatures are lower; these are the chill factors. Evaporative cooling is the loss of heat through perspiration and contact with the air. Evaporative cooling also occurs in the lungs in the breathing process. Decreasing humidity and higher wind speeds will increase evaporative cooling (Herrington, 1973).

Cities tend to be warmer than the surrounding countryside on an average of 0.5 to 1.5°C (Federer, 1970). This difference may be desirable in the winter, but in the summer it can cause discomfort. The summer difference is mainly because of the lack of vegetation in the city and its role in the absorption of solar radiation and in evaporative cooling.

As solar radiation enters the earth's atmosphere, part of it is lost through cloud-cover reflection; some of it is scattered and diffused by particles in the atmosphere; some is absorbed by gaseous pollutants (e.g., carbon dioxide, water vapor, and ozone); the remainder (approximately one-half) penetrates to the earth's surface.

During the daylight hours, solar radiation is absorbed by city surfaces—asphalt, concrete, steel, glass, tar roofs, and others. All of these are poor insulators—gaining, but also losing heat more readily than vegetation or soil. Thus, there is usually a considerable temperature differential between these surfaces and the surrounding air. Heat is then either convectively transferred to the air, causing an increase in surrounding air temperatures or is conducted to subsurface materials. Usually, the resulting increase in air temperature will also bring about a decrease in relative humidity.

Trees, shrubs, and grass ameliorate air temperatures in urban environments by controlling solar radiation. Tree leaves intercept, reflect, absorb, and transmit solar radiation (Figure 4.1). Their effectiveness depends on, for example, the density of species foliage, leaf shape, and branching patterns. Deciduous trees are very instrumental in heat control in urban settings in temperate regions. During the summer they intercept solar radiation and lower temperatures. In the winter the loss of their leaves results in the pleasant warming effects of increased solar radiation (Figure 4.2).

Trees and other vegetation also aid in ameliorating summer air temperatures through evapotranspiration. Trees have been called nature's air conditioners. A single isolated tree may transpire approximately 88 gallons (400 liters) of water per day (provided that sufficient soil moisture is available) (Kramer and Kozlowski, 1970). This has been compared to five average room air conditioners, each with a capacity of 2500 kcal/hr, running 20 hours

51

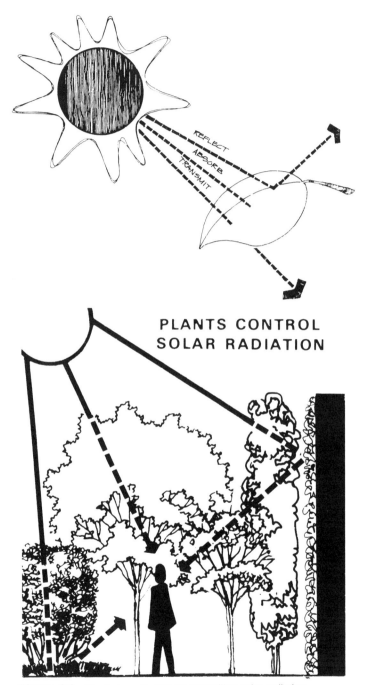

REFLECT

ABSORB

TRANSMIT

**PLANTS CONTROL
SOLAR RADIATION**

Figure 4.1 Plants benefit man by controlling solar radiation.

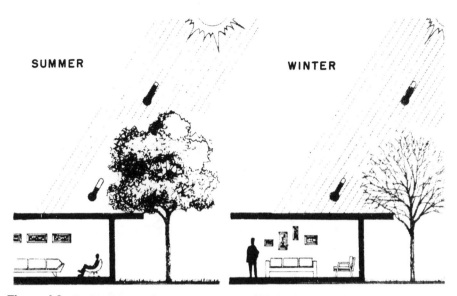

Figure 4.2 Deciduous trees intercept the sun's rays during the summer months and allow them to pass through during the winter.

a day (Federer, 1970). In a forest situation on a calm day, temperature decreases and relative humidity increases downward through the canopy profile (Figure 4.3).

For building cooling, Parker (1983) has outlined the concept of "precision landscaping," involving the analysis of climatic data of a specific locale combined with the termal and energy utilization patterns of particular buildings. Fundamental to this concept is that landscaping reduces building cooling requirements in three ways: (1) blocking solar radiation from the building, the adjacent ground, and the foundation; (2) creating cool microclimates near the building by evapotranspiration; and (3) either channeling or blocking air flows through and around the building. Thus, by precisely locating landscape plants so that these effects can be maximized during seasonal peak demand periods for electricity, considerable savings can be realized. Parker's studies show that west wall temperatures can be reduced by 28°F on very hot, humid afternoons in south Florida by large-canopied trees on the west and hedges adjacent to walls. His work also emphasizes the importance of shading outside air-conditioning units—the strategic location of only one or two small trees can increase operating efficiency by as much as 10 percent.

Although not specifically in urban sites, it was found in Pennsylvania that the cooling energy needs of mobile homes in a deciduous forest were

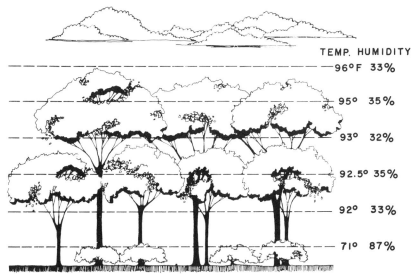

TEMP.	HUMIDITY
96°F	33%
95°	35%
93°	32%
92.5°	35%
92°	33%
71°	87%

Figure 4.3 Temperature decreases and humidity increases downward through a forest canopy.

CANOPY EFFECT—NIGHT

Figure 4.4 A tree canopy will slow the loss of heat at night.

75 percent less that in the open (DeWalle, Heisler, and Jacobs, 1983). In the same stand in winter, however, shade (from deciduous trees) offset savings from the reduced infiltration of cold air (wind protection), resulting in net heating energy savings of only eight percent. In the dense winter shade of a pine forest, heating energy needs rose 12 percent.

At night heat is lost primarily through the exchange of infrared radiation between city surfaces and the atmosphere. On cool, clear nights, city surfaces cool more rapidly; on cloudy nights, there is less cooling. In addition, the rate of infrared heat loss varies with the type of material originally receiving heat from solar radiation during the day. Heavy, high density materials cool slowly; thus releasing heat that may be desirable. At night, tree canopies slow the loss of heat from city surfaces, providing a screen between the cooler night air and the warm surface materials (Figure 4.4). Thus, the night temperatures are higher under trees than in the open (Federer, 1976). In urban areas, this differential may often be as great as 10 to 15°F (5 to 8°C) (Federer, 1970).

Another major influence on heat loss is the actual physical makeup of the city. A city is not a uniform object; it has many different parts and many different structures, all of which work together in varying ways: for example, narrow streets, tall buildings, factories, low buildings, wide streets, parking lots, parks, courtyards, lakes, hills, and rivers among others. Each location within a city has its own microclimate and is unique in its effect on the inhabitants and their sense of comfort. These microclimatic effects will be further discussed in Chapter 5.

Wind protection and air movement

Air movement, or wind, also affects human comfort. The effect may be either positive or negative depending largely on the presence or absence of urban vegetation. Wind can increase evaporative cooling during the day (Figure 4.5). The cooling will vary with the surrounding terrain and the wind speed (Federer, 1970).

During summer, air movement has relatively little effect on air temperature unless the wind is part of a cool front. Instead, it simply causes a cooling sensation bcause of convective heat loss and evaporative cooling. In Figure 4.3, it is shown that because trees screen sunlight and transpire moisture, the area below the forest canopy can be as much as 25°F (14°C) cooler on a still summer day than in an open area. Wind can reduce this temperature differential by replacing moist, cool air with dry, warm air. Hot, dry foehn or chinook winds can actually cause heating. Winds themselves can also be caused by temperature differences. Warm air rising over cities adjacent to oceans or lakes may pull in low, cool breezes; warm air rising from hot, dry open areas may pull in cool air; or cool air may flow downhill into low areas.

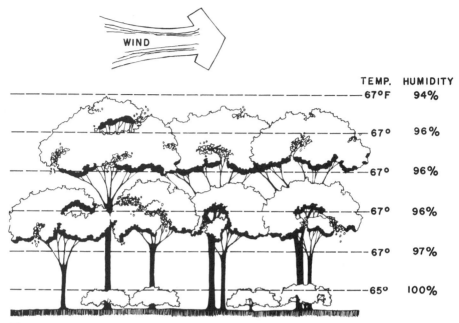

Figure 4.5 Wind can negate cooling effects normally found within a forest canopy.

Trees reduce wind velocity and create sheltered zones both leeward and windward. Thus, trees can interfere with the evaporative cooling process, allowing higher temperatures to prevail in protected zones. This effect is greater from dense coniferous plantings than deciduous tree and shrub plantings and may be positive or negative depending on the time of the year. In the winter, a dense row of coniferous trees planted next to the north and west walls of a home in an area where prevailing winds come from these directions may create a "dead-air space" or insulation zone that prevents heat loss from a building (Figure 4.6). With a constant 70°F (22°C) house temperature, this could result in a fuel savings of nearly 23 percent (Robinette, 1972).

Conifers planted on slopes can impede the movement of cold air that would normally flow to low points, thus preventing the formation of localized frost pockets.

In many parts of the world, trees are used as effective wind controllers. This is particularly true in the Great Plains and other relatively treeless areas where severe winds cause discomfort, irritability, and occasional loss of life and property. In these regions, a belt of trees to slow the wind is a welcome relief (Figure 4.7).

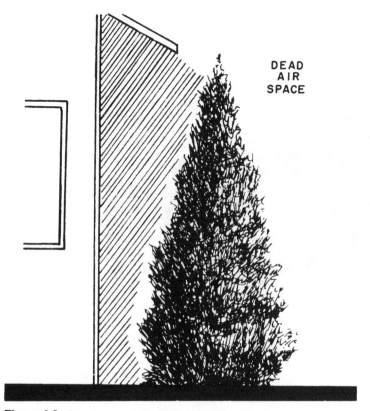

DEAD
AIR
SPACE

Figure 4.6 The placement of coniferous trees next to a house can reduce heat loss during winter.

Before discussing how trees are used in controlling wind, it is necessary to understand wind movement. Two types of wind or airflow are generally recognized:

1. Laminar airflow: layers or streams of air that flow one on top of another.
2. Turbulent airflow: air masses moving in the same direction but in a random pattern.

Turbulence in airflow is governed by disturbances in the airstream and roughness of the surfaces over which it flows. The more streamlined the surface obstruction, the less the turbulence. Layer separation can occur when air moves around sharp corners. Usually wind without obstruction flows in parallel layers (laminar flow). When wind flows over an obstructive surface, the layer of air adjacent to the surface—the boundary layer—usually

57

Figure 4.7 One of many shelterbelts in the Great Plains that are effective in reducing wind movement.

speeds up, creating a low pressure area between this boundary and the surface obstruction. Such low pressure will either result in the movement of the obstruction or will tend to pull the boundary layer back into its original position. This process is affected by the intensity of the wind speed and shape of the obstruction. Usually the leeside of the surface barrier is protected (air movement is slowed). This leeward area of protection decreases as wind speed increases. The protected area on the leeward side increases as steepness of the slope on the windward side of the barrier increases. In addition, pierced and incomplete barriers may actually increase comfort in the leeward zone as they prevent rapid return of the original wind speed for a greater distance on the leeside.

Trees and shrubs control wind by obstruction, guidance, deflection, and filtration. Effect and degree of control vary with species size, shape, foliage density and retention, and the actual placement of the plants. Obstruction involves the placement of trees to reduce wind speed by increasing resistance

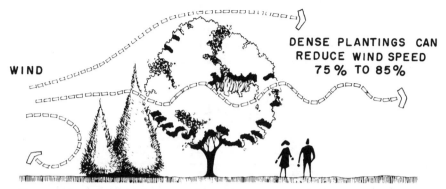

DENSE PLANTINGS CAN
REDUCE WIND SPEED
75% TO 85%

WIND

Figure 4.8 Trees aid human comfort by reducing windspeeds.

to wind flow (Figure 4.8). An example is the prevention of heat loss by dense rows of conifers adjacent to a building, as mentioned earlier. Trees, by themselves or in combination with other barriers, may alter airflow over landscapes and around buildings. Plants can be situated to eliminate wind currents around corners or at entrances to buildings. Careful placement is necessary, however, as plants can also obstruct desired air movement through buildings.

Deflection of wind by trees and shrubs has been a topic of study since the 1930s in the Great Plains. Windbreaks perpendicular to prevailing winds may reduce wind two to five times the height of the tallest trees in front of the barrier and for distances of 30 to 40 times on the leeward side (Figure 4.9). Maximum wind reduction occurs in leeward distances of 10 to 20 times the height of the tallest trees. Here, downwind velocities may be reduced as much as 50 percent. The actual degree of protection depends on the height, width, penetrability, row arrangement, and species in the windbreak.

The wake zone depends largely on tree height; the taller the trees, the greater the protected distance. As trees increase in height, they generally

H = AVERAGE WIND BREAK HEIGHT

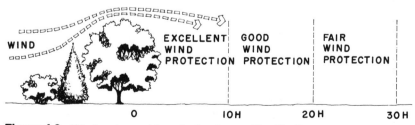

WIND

EXCELLENT WIND PROTECTION | GOOD WIND PROTECTION | FAIR WIND PROTECTION

0   10H   20H   30H

Figure 4.9 Windbreaks provide protection for considerable distances.

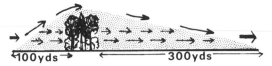

A 30ft. high shelterbelt affects wind speed for 100yds.in front of the trees and 300yds. down wind.

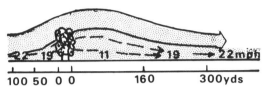

Effect of moderately penetrable wind-breaks on wind.

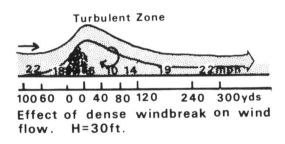

Effect of dense windbreak on wind flow.   H=30ft.

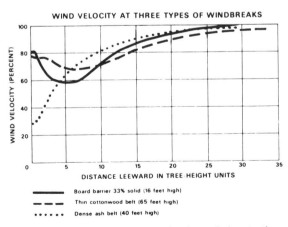

Figure 4.10 Density affects the downwind protection zone of any given windbreak.

become more open and allow increased airflow in the lower portions. Here more rows of trees in combination with low shrubs are usually required to provide adequate protection. Thus, the effectiveness of a windbreak depends on both height and penetrability. Usually the width of a windbreak has a negligible effect in reducing wind velocity on the leeward side; it can, however, create a noticeable microclimate within the sheltered area. Windbreaks with a cross-section design of a pitched roof are less effective in blocking winds than those with vertical edges.

While very dense windbreaks provide a greater wind reduction on the leeward side, the total zone of effective shelter is less. A penetrable wind barrier may not provide as much wind reduction on the leeward side, but the wake zone extends for a greater distance (Figure 4.10). Movement of air through the windbreak gives some lift to the boundary layer and acceleration to normal wind speed is more gradual. The optimum density for a shelterbelt is estimated at 50 to 60 percent (Robinette, 1972).

Species selection is important in windbreak efficiency. Conifers with dense foliage are best on the north and west sides where protection from winter winds is desired. Deciduous species are preferred on the south and east as they protect against hot, dry winds in the summer and allow incoming solar radiation in winter.

Trees can provide effective wind protection for roads and highways. Highways perpendicular to prevailing winds are often subjected to strong crosswinds and gusts. Wind movement is also affected by highway cuts, breaks in terrain, and nearby buildings. Wind movement can often be controlled in such problem areas by the proper placement of trees and shrubs. (Figure 4.11).

Trees can also be used to control snow drifting. As wind velocity is slowed to the leeward, snow particles are deposited. Windbreaks containing low shrub rows are most effective in the control of drifting. Drift patterns vary with windbreak density (Figure 4.12). Dense windbreaks yield drifts that are narrow and quite deep. As penetrability increases, drifts become wider and shallower. Higher velocity winds cause drifts to be narrow and deeper. The following formula has been derived for computing windbreak drift patterns (Jensen, 1954):

$$L = \frac{36 + 5h}{K}$$

where: L = the length of the drift in feet
h = the height of the screen in feet
K = the function of the screen density, 1.0 for 50 percent density and 1.28 for 70 percent density

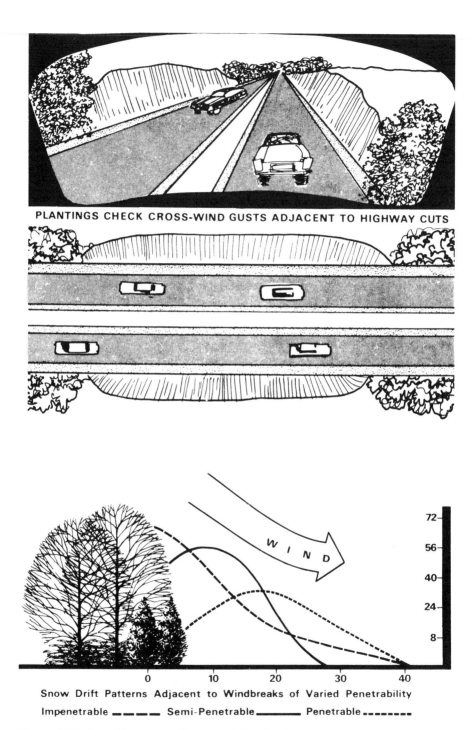

PLANTINGS CHECK CROSS-WIND GUSTS ADJACENT TO HIGHWAY CUTS

Snow Drift Patterns Adjacent to Windbreaks of Varied Penetrability

Impenetrable _ _ _ _   Semi-Penetrable _____   Penetrable _ _ _ _ _ _ _

Figure 4.12 Snowdrift patterns adjacent to windbreaks of varying density.

Windbreaks may be more desirable than snowfences since maintenance costs of snowfencing can be high, and certainly trees are more esthetically pleasing. Trees and shrubs can be used to provide drift-free parking lots, roads, and sidewalks (Figure 4.13). In addition, they may be used to cause snow deposition where desired, such as ski areas, toboggan runs, and where soil moisture from melted snow is desired.

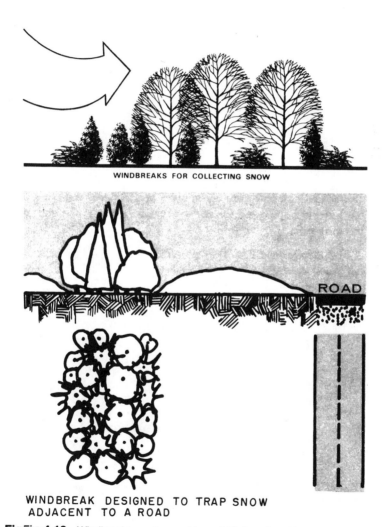

WINDBREAKS FOR COLLECTING SNOW

ROAD

WINDBREAK DESIGNED TO TRAP SNOW ADJACENT TO A ROAD

Figure 4.13 Windbreaks can be used to assist in keeping sidewalks and roads free of snow drifts.

63

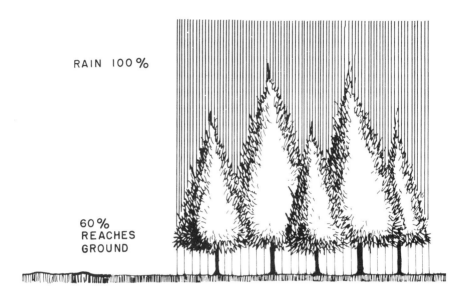

RAIN 100%

60%
REACHES
GROUND

Pine forest.

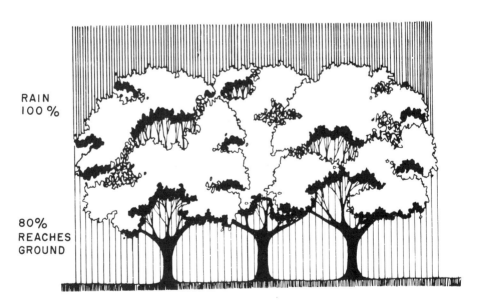

RAIN
100%

80%
REACHES
GROUND

Hardwood forest.

Figure 4.14 Coniferous forest trees intercept more rainfall than do hardwood forest trees.

Precipitation and humidity

Trees intercept and filter solar radiation, inhibit wind flow, transpire water, and reduce evaporation of soil moisture. Thus, beneath a forest canopy, humidity is usually higher and evaporation rates are lower. Temperature beneath the canopy is also lower than the surrounding air during the day and warmer during the evening.

In addition to their effects on temperature, trees and shrubs are important in the hydrologic cycle. They intercept precipitation and slow its descent to the soil surface. This can increase infiltration and decrease runoff and soil erosion. Trees may also reduce soil moisture evaporation. However, their high transpiration rates and interception process may actually reduce the amount of water available for aquifer recharge or reduce stream flow when compared to other types of vegetative cover. All of these factors are important in urban areas where natural underground aquifers are the main source of water.

Effectiveness in controlling runoff and increasing infiltration varies with soil type, organic content of the soil, topography, type, and intensity of precipitation, and composition of vegetative cover. Interception of precipitation by coniferous trees is usually greater than that of hardwoods (Figure 4.14). An estimated 60 percent of the rainfall will reach the ground through a pine canopy as compared to 80 percent through a hardwood canopy. This is because the leaf structure of conifers enables better entrapment of water droplets (Figure 4.15). Pubescence is another leaf characteristic important in water entrapment. Branching patterns also affect interception rates, horizontal branching patterns being the most effective (Figure 4.15). In addition, rough bark slows the movement of water down tree trunks (Figure 4.15). The more intense the rainfall, however, the less effective a tree will be in interception. The same is true of duration of rainfall. Resultant runoff will also be affected by the nature of the shrub canopy, ground cover, litter, and topography.

Engineering Uses

In recent years, highly specialized uses for plants in solving environmental engineering problems have been developed. Involved are not only landscape esthetics but soil erosion control, air pollution, noise abatement, wastewater management, traffic control, and glare and reflection reduction. Robinette (1972) has listed the following plant characteristics and their effects that help to solve environmental engineering problems:

1. Fleshy leaves that deaden sound.
2. Branches that move and vibrate to absorb and mask sounds.

65

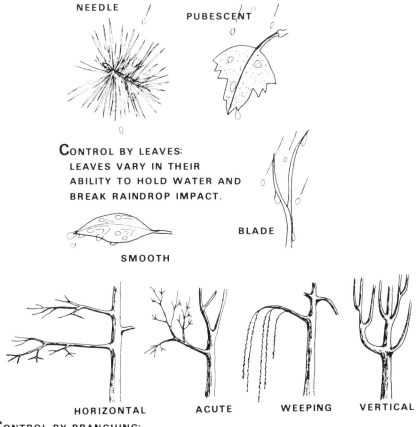

NEEDLE PUBESCENT

CONTROL BY LEAVES:
LEAVES VARY IN THEIR
ABILITY TO HOLD WATER AND
BREAK RAINDROP IMPACT.

BLADE

SMOOTH

HORIZONTAL ACUTE WEEPING VERTICAL

CONTROL BY BRANCHING:
HORIZONTAL BRANCHING IS MOST EFFECTIVE IN PREVENTING WATER
RUNOFF DOWN TREE TRUNK AND EROSION AT BASE OF TREE.

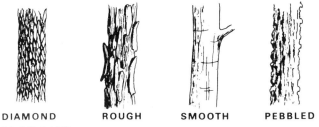

DIAMOND ROUGH SMOOTH PEBBLED

CONTROL BY BARK:
ROUGH BARK HOLDS AND SLOWS WATER RUNNING DOWN
TRUNK AND PREVENTS EROSION AT BASE OF TREE.

Figure 4.15 Differences in plant species morphology will affect how much
rainfall is intercepted.

66

3. Pubescence on the leaves to entrap and hold dust particles.
4. Stomata in the leaves to exchange gases.
5. Blossoms and foliage that provide pleasant smells to mask odor.
6. Leaves and branches to slow wind.
7. Leaves and branches to slow rainfall.
8. Spreading roots to hold soil against erosion.
9. Dense foliage to block light.
10. Light foliage to filter light.
11. Spiny branches to deter human movement.

Erosion control and watershed protection

Because of environmental impacts associated with most construction activities, erosion control is perhaps the most important engineering use of plants. Soil erosion is the loss of topsoil by wind or water movement, usually resulting from improper soil protection. Soil erosion is influenced by exposure of the site to wind and water, the physical characteristics of the soil, and topography.

Wind erosion in the urban forest is not as well recognized as water erosion. Wind erosion is mostly associated with rural agricultural lands where large land surfaces may lie unprotected, where winds are frequent and strong, and where precipitation is minimal and unpredictable. Visibility is often reduced, presenting a safety hazard. Wind erosion is affected by speed, duration, and direction of the wind along with such soil factors as moisture, physical structure, and cover. Normally, as the wind moves over exposed dry soil, the smaller soil particles are moved. This results in saltation, a condition where larger particles moving along the soil surface act as abrasives, causing smaller particles to dislodge and become windborne. Trees and shrubs have long been used to reduce wind erosion. During the 1930s, thousands of windbreaks were established in the Great Plains for this purpose, and new ones are still being planted.

Soil erosion in urban areas is usually associated with construction activities where soil surfaces have been exposed. Splash erosion from the impact of falling raindrops causes soil particles to be dislodged and to move into runoff suspension where they act as scouring agents removing even more soil.

There are four types of runoff erosion: (1) sheet, (2) rill, (3) gulley, and (4) slip. Sheet erosion is the removal of the entire top soil layer from an exposed site. As this continues, soft areas in the soil surface may be removed faster, resulting in rills or troughs. Channeling of runoff water in rills results in gulley erosion. Slip erosion is the mass movement of water-saturated soils downslope. The latter is associated with steep slopes in regions of high rainfall following vegetation removal by fire or other causes.

Drastic changes occur in watershed hydrology when agricultural or forested areas are converted to urban land use. The most important thing city planners can do to minimize adverse hydrologic effects of urbanization is to retain partial watersheds. This entails reserving areas along streams, around lakes, or other areas as greenbelts to be maintained in their natural state (Figure 4.16). Other measures that can improve the hydrology of urban areas include:

1. Perforation of compacted lawns to improve water absorption and infiltration.

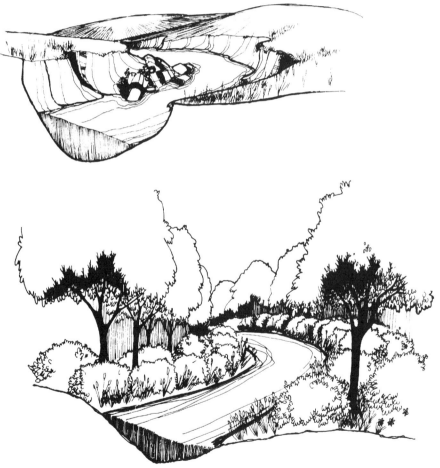

Figure 4.16 Retaining of forested areas adjacent to streams not only reduces erosion but is also more esthetic.

2. Planting of trees and shrubbery.
3. Use of conservation practices such as terracing, contour planting, mulching, and sod waterways in yards and gardens.
4. Construction of gravel drainage areas for roof runoff.
5. Use of flat roofs to reduce runoff and aid building cooling.
6. Temporary mulching of construction sites.

Plants reduce water-caused soil erosion by intercepting rainfall, by holding soil with their roots, and by increasing water absorption through the incorporation of organic matter (Figure 4.17). In addition, plants are more attractive than mechanical water erosion control devices.

Management of watersheds in and adjacent to urban areas is highly complex in that it must be carried out in an environment of physical factors, economics, and politics. Management must ultimately involve the manipulation of vegetation and other physical factors: to (1) influence water yields and (2) to maintain or improve water quality. Lynch (1983) indicated the complexity and difficulty of watershed management with the thought that while research has shown that water yield from forested areas increases as evaporative losses are reduced, and maximum water yields are therefore attainable from clearcutting, he could not conceive of any situation (at least in the East) in which an urban forest watershed would be completely harvested for the sole purpose of increasing stream discharge. Other values of the forested area are simply too great—timber, wildlife, recreation, and esthetics. Thus, multiple use is the only sensible approach, and as with all forested areas, there must be a primary resource. In this case, it is water.

Recent federal and state legislation authorizes (regulates) forest management for water quality. Of particular concern are forest harvesting operations that contribute to nonpoint pollution, resulting in increased turbidity, suspended sediment, and increased water temperature. Also of concern are other forestry operations involving the use of pesticides and fertilizers. Other uses of forested watersheds also influence water quality, particularly recreation, with the physical impact of people and their refuse. In spite of these problems, as Lynch (1983) admonishes, the alternative of "no management" is unacceptably costly. It becomes the urban forest watershed manager's responsibility then to supply water "in a manner that maintains a quality environment where esthetic, cultural, recreational, and economic values are concerned."

Wastewater management

Rapid population growth accompanied by an expanding industrial base has greatly increased the demands on our nation's water resources. This growth is also creating ever-increasing waste disposal problems. In 1970 approxi-

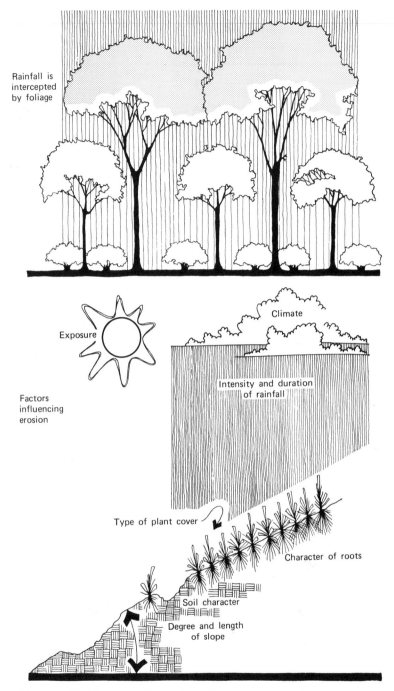

Figure 4.17 The various factors that influence water-related soil erosion.

Rainfall is intercepted by foliage

Exposure

Climate

Intensity and duration of rainfall

Factors influencing erosion

Type of plant cover

Character of roots

Soil character

Degree and length of slope

mately 25 billion gallons (113 billion liters) of municipal sewage effluent were discharged daily into surface streams in the United States. It is estimated that the population of the United States will reach 300 million by the year 2000 A.D. and the municipal waste load of the nation would increase correspondingly (Nutter, 1971).

Wastewater disposal begins with "primary treatment." Large objects are removed and the remaining material is ground in a grit chamber and piped to sedimentation tanks where sand, silt, and suspended organic material are allowed to settle. After the water is skimmed, it is discharged into lakes or streams. At this stage about one-third of the biochemical oxygen demand (B.O.D.)* still remains. Only a very small portion of the nutrients, however, have been removed. In about 30 percent of the United States' treatment plants, this is all that is done.

Secondary treatment is primarily biological in nature and involves the use of microorganisms (bacteria) to decompose organic compounds. Approximately 60 percent of the treatment plants in the United States incorporate this step (Owen, 1975). In this process, about 90 percent of the B.O.D. and 90 percent of the suspended solids are removed. However, 58 percent of the nitrogen compounds and 30 percent of the phosphorus compounds (elements responsible for stream eutrophication) still remain. Discharge of this nutrient-rich secondary effluent into lakes or streams causes abundant aquatic growth. These plants grow profusely, die, decompose, and increase the B.O.D. of the water. This accelerated process utilizes available dissolved oxygen that is necessary for other aquatic life. Nutrients are re-released to be recycled in the same manner. As a result, streams may become even more polluted after receiving secondary treatment water than they were when receiving raw sewage.

Recently a third process, called a tertiary treatment, has been developed to aid in nutrient removal. In this process secondary effluent is mixed with lime to remove phosphorus by precipitation. Nitrogen, present primarily as ammonia, is removed by blowing large amounts of air through the effluent. Though highly effective, these treatments are expensive—30¢ per 1,000 gallons (4,545 liters) of effluent as compared to 12¢ per 1,000 gallons (4,545 liters) for secondary treatment. Thus, tertiary treatment may not be feasible in many urban and industrial areas.

Because of the high cost of current methods, other solutions for effluent disposal should be considered. A promising alternative is land sewage disposal. Land disposal systems reduce stream water pollution, conserve and recycle water, and allow nutrients to be recycled for further use. The suit-

---

*B.O.D. is a measure of the oxygen consumed during decomposition of organic compounds.

ability of effluent for land disposal depends on its chemical composition. Wastewater from residential areas can generally be used for this purpose whereas industrial wastewater may require some treatment to remove toxic substances that could affect biological systems. Industrial waste from food processing plants and pulp mills are usually well-suited for land disposal. Land disposal is not a recent innovation. Most of the early systems, however, were concerned only with waste disposal and did not consider water conservation or the recycling of nutrients.

The following case is made for productive land disposal systems by Sopper (1971):

A more feasible method might be to apply the wastewater to the land so as to utilize the entire biosystem—soil and vegetation—as a "living filter" to renovate the effluent for groundwater recharge. Under controlled application rates to maintain aerobic conditions within the soil, the mineral nutrients and detergent residual might be removed and degraded by micro-organisms in the surface soil horizons, chemical precipitation, ion exchange, biological transformation, and biological absorption through the root systems of the vegetative cover. The utilization of higher plants as an integral part of the system to complement the microbiological and physiochemical systems in the soil is an essential component of the living-filter concept and provides maximum renovation capacity and durability to the system.

Land sewage disposal has been used for centuries in the Orient, and many European cities have had operational systems for years. Systems in Berlin and Paris date back to 1850. Israel has adopted this type of sewage disposal system because of water shortages. The sewage of Tel Aviv has been used to grow fruits and vegetables in the Negev Desert instead of polluting the beaches of the Mediterranean.

In the United States, sewage was used for crop irrigation only on a small scale in the nineteenth and early twentieth centuries, and until recently the process was relatively unknown. The primary reasons for this were:

1. Low population densities and greater abundance of land and water resources.
2. Availability of inexpensive chemical fertilizers.
3. Generally negative public attitude about eating food that was produced from human effluent.

However, these factors have changed in recent years, and agronomists, horticulturists, ecologists, and others are now investigating the use of renovated wastewater for fertilization and groundwater recharge. Industrial wastes have been used for irrigation purposes. For example, at Seabrook Farms in Bridgeton, New Jersey the 10 million gallons (45 million liters) of water used daily to wash vegetables are afterwards disposed of on forest land. Other companies have initiated similar systems, and many paper and pulp mills have incorporated land disposal systems into their operations.

The Pennsylvania State University Waste Water Renovation and Conservation Project is another example (Sopper, 1971). Water pollution and supply were problems in State College. Sewage water was discharged into the area's main watercourse, depleting the stream of necessary oxygen for aquatic life and reducing its recreational value. The main water supply came from groundwater reservoirs that were being depleted by a drought of nine years duration. Millions of gallons of water were released into the stream, thus removing it from the water supply.

The need to solve both the water quality and the water supply problems prompted studies involving the land disposal of wastewater and the living filter concept. The removal of the effluent from local streams to overland disposal sites would reduce stream pollution and enhance renovation and recharge of the groundwater reservoir. A team of scientists studied changes in plant community structure, monitored foliar and soil nutrient changes and absorptive capacities of soils at various levels of effluent application. Their work indicates that the living filter concept is feasible given the following site conditions (Sopper, 1971):

1. The soil must have an infiltration and percolation capacity high enough to accommodate applications of wastewater at the recommended rates. Soil permeability must be high enough to permit drainage of the renovated effluent and to maintain aerobic soil conditions. Forest soils are well-suited because they are usually very porous, and have high infiltration and percolation rates, and remain so during the winter season.
2. The soil must have sufficient chemical absorptive capacity, water retentive capacity and depth to the groundwater table to hold dissolved minerals for use by plants and micro-organisms, thereby preventing groundwater contamination.
3. Land must have low relief and good vegetative cover along with accumulated surface organic matter to minimize surface runoff. This is particularly true in the winter months.
4. There must be a groundwater aquifer with a fairly deep water table to accommodate subsequent changes in groundwater storage, and with sufficient horizontal permeability to allow for adequate lateral displacement of the renovated wastewater.

73

Renovation and reuse of wastewater is more widespread in the United States than one might imagine. At Whittier Narrows in Los Angeles County, California, 10 to 15 million gallons (45 to 68 million liters) of sewage waste per day are recovered, purified, and diluted to recharge groundwater supplies for municipal reuse. At Santee, California municipal wastewater is reclaimed and used in a series of recreational lakes. Chanute, Kansas during a severe drought in 1957 impounded its own wastewater, treated it, and reused it for several months. The Desert Inn Hotel and Country Club in Las Vegas, Nevada uses purified sewage effluent to irrigate its 130-acre (52-ha) golf course.

According to the Pennsylvania State University study, the living filter system is best adapted to small cities and suburbs because of the availability of open land close to a treatment facility. In their studies, they found that an application rate of 2 in. (5.08 cm) per week required only 129 acres (52 ha) of land to dispose of one million gallons (4.5 million liters) of wastewater per day. It is estimated that a city with a population of 100,000, discharging 10 million gallons (4.5 million liters) of wastewater daily, would require only 1,290 acres (522 ha) for a living filter system.

Although large blocks of agricultural and forest land may be used for disposal, cities might also use golf courses, recreation fields, forest preserves, parks, greenbelts, scenic parkways, and perhaps even highway medians.

Land treatment approaches can be used in a variety of ways to solve both land use and water quality problems. They can be initiated to create interesting urban forestry projects. Most, if not all, land treatment systems can be designed to use trees or shrubs as a major crop or landscape feature. Varieties of trees to be planted will, of course, depend on the amount of water passed through the soil filter.

It is clear that land can be acquired to build urban forests—parks and other recreation areas, open spaces, and cropped forests—into the urban infrastructure. It appears that land must also be acquired for waste water treatment using trees. In fact, unless state and local governments intend to lower water quality standards, thereby endangering water supplies and denying citizens the use of streams and lakes for water contact recreation, land treatment will have to be used to treat excessive pollutant discharges from malfunctioning conventional sewage treatment plants. Uptake into living root systems appears to be the only feasible method currently known to remove nutrients effectively.

Noise abatement

Noise is commonly thought of as excessive or unwanted sound. Experts in the field often refer to noise as "invisible pollution." Noise involves both physical and psychological effects. The physical effects deal with transmis-

74

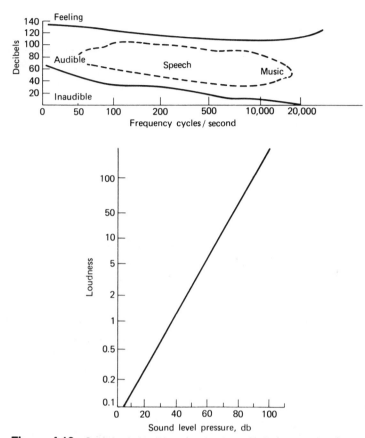

Figure 4.18 Sound relationships showing how decibels are related to sound frequency and loudness.

sion of sound waves through the air, while the psychological effects involve the human response to sound. Sound is defined as a pressure variation in an elastic medium that affects the hearing mechanism. Sound moves in waves; the high pitches or frequencies have shorter wavelengths and the low frequencies longer wavelengths. Sound frequency is measured in cycles per second (CPS). The human ear can detect sound in the range of 20 to 20,000 CPS. The sound pressure level (SPL) or intensity of sound is measured in decibels (dB). The decibel scale, when adjusted by weighting the sound level meter to the dBA scale is approximately related to loudness, but that relationship is not intended (Figure 4.18). The lowest sound that can be detected by a keen human ear under very quiet conditions corresponds to zero decibels, and the highest sound corresponds to about 120 decibels

TABLE 4.1
Sound Levels in dBA (from Herrington, 1973)

Source	SPL, dBA
Threshold of pain	130
Jet takeoff (200 ft away)	120
Large truck, noisy motorcycle or pneumatic drill (50 ft away)	80
Vacuum cleaner (10 ft away)	70
Average residence	50
Residential area, night	40
Soft whisper (5 ft away)	30
Threshold of hearing	0

(Table 4.1). The decibel scale is logarithmic, and is defined by an equation.* Because of the logarithmic nature of the decibel and the response characteristics of the ear, an increase of 10 dB corresponds to the approximate doubling of the loudness of a sound.

Propagation of outdoor noise is caused by: (1) the nature of the source (e.g., its frequency, composition, location, and whether or not the source is of a linear or pooint nature); (2) the nature of the terrain and vegetation over which the sound is passing; and (3) the state of the atmosphere (e.g., speed and direction of the wind and temperature conditions).

Outdoor sounds are normally attenuated (reduced in intensity) before reaching the receiver. Attenuation usually assumes a dual nature; it is called *normal* attenuation when it is related to distance and *excess* attenuation when it results from the introduction of elements or barriers between the sound source and the receiver.

Noise originating from a point source (as from a mountain top) radiates spherically away from the source. Acoustic energy decreases in proportion to the reciprocal of the square of the distance. Studies have shown that each time the distance from a point noise source is doubled, there is a reduction of 6 dBA. Thus, if a truck produced an SPL of 80 dBA at roadside [10 ft (3.1 m) from the truck], the SPL at 20 ft (6.3 m) would be 74 dBA; at 40 ft (12.5 m), 68 dBA; at 80 ft (25 m), 62 dBA; etc. However, with a noise coming from a line source (a busy freeway), the reduction is only 3 dBA for each doubling of the distance.

Excess attenuation of sound propagated along the ground is greatly affected by the presence of temperature and wind gradients. Attenuation

---

*SPL (decibels $= 20 \log p/P^0$, where p is measured pressure and P^0 is a reference pressure of 0.0002 microbars or 0.00002 Newtons per square meter.

measured upwind may exceed those measured downwind by as much as 25 to 30 decibels. In general, sounds passing downwind from a source are directed downward toward the surface, while sounds traveling upwind from a source are directed upward from the surface. At night, when the atmosphere is coldest near the ground, sound waves will be directed toward the surface, while during the day when the atmosphere is warmer near the ground, sound will be directed away from the surface.

Elements introduced between the source and the receiver reduce sound by absorption, deflection, reflection, and defraction. Another phenomenon, masking, may also aid in reducing noise.

Absorption occurs when an object receives sound waves and entraps them, converting the sound into other energy forms and, eventually, heat. If the material is sufficiently porous, up to 95 percent of the energy may be absorbed. *Deflection* occurs by introducing an element that will cause the noise to bounce from a recipient into an area of direction less offensive to the hearer. *Reflection* causes the sound to be directed back to the source, thus protecting the hearer. *Refraction* is the dissipation, diffusion, or disperson of acoustical energy when it strikes a rough surface. This may also be caused by wind turbulence. Masking, though not a form of attenuation, is the introduction of a pleasing sound that tends to override undesirable noises (e.g., piped-in music in an office).

Research documentation of plants as attenuators is limited. Robinette (1972) lists three early works in his discussion of plants and acoustical control: Eyring (1946); Weiner and Keast (1959); and Embelton (1963). The most recent studies have been by Cook and Van Havrbeke in Nebraska (1971, 1974, 1977).

How effectively plants control sound levels is determined by the sound itself, the planting involved, and the climatic conditions. The properties of sound include type, origin, decibel level, and intensity. Planting factors include the species, arrangement in relation to the sound source and the receiver, and planting height and density (Figures 4.19 and 4.20). Climatic factors include wind direction and velocity, temperature, and humidity. In addition to deflection, absorption, refraction, reflection, and masking, plants may also attenuate sound by their influence on local climate; that is, their stabilization of temperature, their modification of wind speed, and other factors. Weiner and Keast (1959) pointed out that sound propagation patterns over woodland areas differ from those over open level terrain. They attributed this difference not only to absorption and scattering, but also to wind velocities and vertical wind gradients, which are reduced over woodland areas. Also air temperatures are fairly uniform because such areas are not subject to the large diurnal lapse rates and nocturnal inversions common to open terrain.

77

NOISE SOURCE CITY TRASH CREW AT A DISTANCE OF 90'
PLANTING CONIFEROUS EVERGREEN HEDGE
NOISE REDUCTION 50%

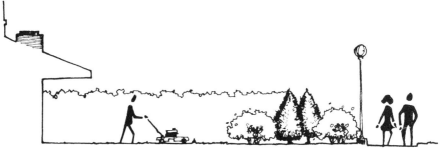

NOISE SOURCE LAWNMOWER EXHAUST AND BLADE WHINE
PLANTING 6' HIGH 10' THICK MIXED CONIFEROUS AND DECIDUOUS
NOISE REDUCTION 40%

NOISE SOURCE CHILDREN PLAYING AT A DISTANCE OF 50'
PLANTING DECIDUOUS PLANTING
NOISE REDUCTION 50%

Figure 4.19 The function of plants in reducing noise pollution in different urban settings.

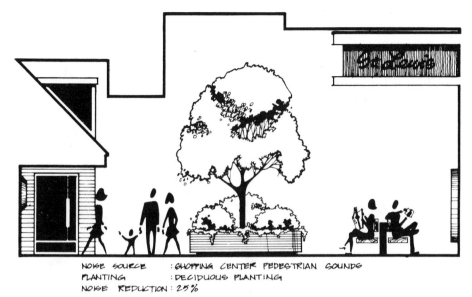

NOISE SOURCE   : SHOPPING CENTER PEDESTRIAN SOUNDS
PLANTING   : DECIDUOUS PLANTING
NOISE REDUCTION : 25%

Figure 4.20 Plants can reduce sound in downtown urban areas.

Plant attenuation of sound is shown in Figure 4.21; sound waves are absorbed by the leaves, branches, and twigs of trees and shrubs. These plant parts are light and flexible. It has been postulated that the most effective plants for absorbing sound are those having many thick, fleshy leaves with petioles. This combination allows for the highest degree of flexibility and vibration (Robinette, 1972). Sound is also deflected and refracted by the heavier branches and the trunks of trees. It has been estimated that, on the average, forests can attenuate sound at the rate of 7 dB per 100 ft (30 m) of distance at frequencies of 1,000 CPS or less (Embleton, 1963).

A recent study involving trees and shrubs as sound attenuators was conducted by the school of Engineering at the University of Nebraska and the Rocky Mountain Forest and Range Experiment Station, U.S. Forest Service. These studies showed that the potential of trees and shrubs for sound reduction is quite promising. Attenuations of 5 to 8 dB and even as much as 10 dB for wide belts of tall, dense trees were recorded. These levels were even greater when harder surfaces were compared with tree-shrub-grass combinations; then attenuations of 8 to 12 dB were common (Cook and Van Haverbeke, 1971). Generally, wide belts of tall trees were found to be most effective, and species did not differ greatly in their ability to reduce noise levels, provided that deciduous trees were in full leaf. Evergreens are desirable for year-round screening (Figure 4.22).

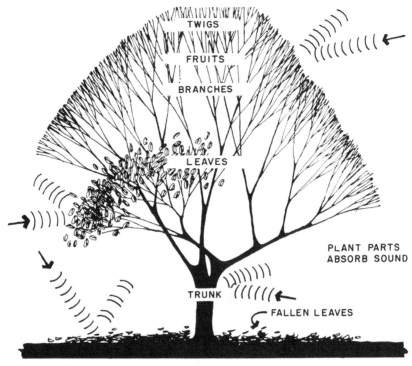

Figure 4.21 A graphic illustration of how plants attenuate sound.

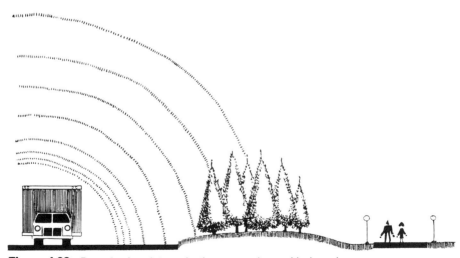

Figure 4.22 Properly placed, tree plantings can reduce vehicular noise in residential areas.

The relative position of the screen between the sound source and the hearer (receiver) is very important. A screen close to the sound source is more effective than one close to the area to be protected.

Urban residential property can be effectively screened from passenger car noise with a single row of dense shrubs backed by a row of taller trees, totaling a depth of 20 feet (6.3 m). Screening for rural areas or freeways requires wider belts consisting of several rows of tall trees in dense plantings. In general, 100 ft. (30 m) or more of such a planting between the noise source and the area to be protected are desirable.

Specific recommendations based on these studies are as follows (Cook and Van Haverbeke, 1971):

1. Reduction of noise from high-speed car and truck traffic in rural areas is best achieved by tree and shrub belts 65 to 100 feet (20 to 30 m) wide with the edge of the belt within 50 to 80 feet (16 to 20 m) of the center of the nearest traffic lane. Center tree rows should be at least 45 feet (14 m) tall.

2. To reduce noise from moderate-speed car traffic in urban areas, tree and shrub belts 20 to 50 feet (6 to 16 m) wide may be used effectively with the edge of the belt from 20 to 50 feet (5 to 16 m) from the center of the nearest traffic lane. Shrubs six to eight feet (2 to 2.5 m) tall should be used next to the traffic lane followed by backup rows of trees 15 to 30 feet (4.5 to 10 m) tall.

3. For optimum results, trees and shrubs should be planted close to the noise source as opposed to close to the protected area.

4. Where possible, use taller varieties of trees which have dense foliage and relatively uniform vertical foliage distribution (or use combinations of shrubs and trees). Where the use of tall trees is restricted, use combinations of shorter shrubs and tall grass or similar soft ground cover, as opposed to paving, crushed rock, or gravel surfaces.

5. Trees and shrubs should be planted as close together as practical to form a continuous, dense barrier. The spacing should conform to established local practices for each species.

6. Evergreens (conifers) or deciduous species which retain their leaves should be used where year-round noise screening is desired.

7. Belts should be approximately twice as long as the distance from the noise source to the receiver, and when used as a noise screen parallel to a roadway, should extend equal distances along the roadway on both sides of the protected area.

Traffic noise can also be alleviated by proper highway design. Studies have been conducted on sound propagation from highways that are depressed, raised, or on a grade. Proper planting in such situations can aid tremendously in sound control. A publication dealing with optimum planting for sound control along highways was published in the early 1950s (Simonson; 1957). This report also discussed planting design for retaining walls and buffer plantings on roadsides to control noise. Figure 4.23 illustrates some of the design combinations for various highway situations.

Where rights-of-way are limited, a combination of wall (masonry, barrier, or embankment) and planting can be used effectively. The effectiveness of the barrier can be increased by the use of vines that cover the wall.

Depressed highways reflect traffic noise upward. Planting on the lower parts of the slope, however, will cause secondary reflection and may reduce effectiveness. Planting along highways is very effective in suppressing noise transmission to adjacent areas.

Preliminary studies indicate that if grass, vines, or other plants are on barrier slopes facing the noise, the noise may be reduced by as much as 8 to 10 dB. Hedges and trees with dense foliage act as sound absorbers and deadeners. The effectiveness of the barrier increases with the increasing thickness, height, and density of the planting.

In work related to the previous discussion Cook and Van Haverbeke (1974) compared bare and tree-covered landforms as attenuators of highway noise. They found that for highway noise abatement, a combination of the two is more effective than either one used separately. The choice is governed by several factors, including cost and availability of material, the time frame involved, and esthetics. The specific recommendations Cook and Van Haverbeke put forth in their study are as follows:

1. To reduce noise from high-speed traffic, land-forms should be sufficient in height to screen the traffic from view. Several rows of trees and shrubs should be planted adjacent to and on the land-form for progressive improvement with conifers used in preference to deciduous species for year-round noise control.
2. The reduction of noise from suburban traffic can be accomplished by the planting of rows of heavy shrubs adjacent to the traffic lanes and constructing a five- to six-foot (two-m) land-form behind the shrubs.
3. Optimum land-form height will vary for each situation; however, 8- to 10-foot (2.5- to 3-m) heights, when used in combinations with taller varieties of trees, are recommended for general applications provided the noise source is screened from view.
4. The noise screen should be placed relatively close to the noise source for maximum benefit and should extend far enough from side to side to assure acceptable sound levels at the protected area.
5. Since wind speed and direction affect sound levels at a specific location, the most severe conditions should be used when estimating expected sound levels for design purposes.

NOISE SOURCE : TRUCK AND AUTO TRAFFIC
PLANTING : DECIDUOUS SHRUBS
NOISE REDUCTION: AUTO 25% TRUCK 50%

NOISE SOURCE : TRUCK AND AUTO TRAFFIC
PLANTING : DECIDUOUS TREES
NOISE REDUCTION: AUTO 20% TRUCK 40%

NOISE SOURCE : TRUCK AND AUTO TRAFFIC
PLANTING : CONIFEROUS
NOISE REDUCTION: AUTO 75% TRUCK 80%

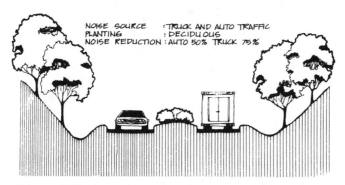

NOISE SOURCE : TRUCK AND AUTO TRAFFIC
PLANTING : DECIDUOUS
NOISE REDUCTION : AUTO 50% TRUCK 75%

Figure 4.23 Sound abatement will vary with the species selected and its placement. In addition roadway design is also a factor.

6. Natural ground configurations, such as hills, ridges, and depressed highways, should be employed to serve as noise screens when planning roadside developments, schools, and residences adjacent to arterial streets in urban areas.
7. Existing trees, shrubs, and grass should be left undisturbed, as far as possible, rather than replacing the soft materials with harder reflecting surfaces detrimental to noise control.

As mentioned earlier, plants can also mask unwanted sounds (Figure 4.24). Plants make their own sounds—the whisper of pines, the rustle of oak leaves, or the quaking of aspens. These are pleasant sounds that tend to make us less aware of more offensive noises. In addition, plants support animals and birds that may make desirable sounds.

Plants can play a role in noise abatement. There are, however, limitations to their effectiveness. Although they may reduce noise to acceptable levels, tree and shrub screens, regardless of their size and density, will not completely eliminate sound. Also, a single tree or a few scattered trees will

SOUND AND PLANTS

PARTS THAT MAKE SOUND

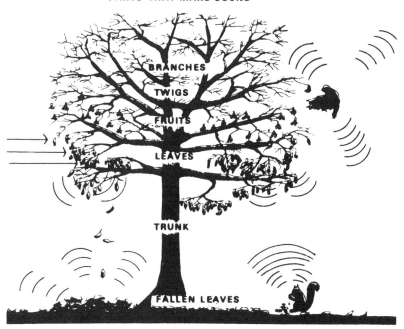

Figure 4.24 Plants can mask noise by creating their own sounds or by enabling accompanying wildlife to do so.

not appreciably reduce noise. Only if trees are massed will they be effective. Trees, shrubs, and other vegetation will aid in noise reduction if used properly, and the advantage will be greater than the psychological aspect of simply screening the noise source from the receiver's view.

Air pollution abatement

The role of trees in reducing air pollution is not well understood, and there is considerable disagreement as to their effectiveness. The types and relative sizes of common air pollutants that affect us are found in Figure 4.25. Most important are those that are either gaseous or particulate.

It is well-known that plants produce oxygen in the photosynthetic process. Some people have suggested that plants perform an important role in reducing air pollution through the processes of oxygenation (the introduction of excess oxygen into the atmosphere) and dilution (the mixing of polluted air with fresh air). Thus, it is reasoned that when polluted air flows in and around plants and through the freshly oxygenated air, dilution occurs and pollution is lessened (Figure 4.26). Granted that these processes occur, are they significant in abating air pollution? Weidensaul (1973) feels that research data do not necessarily support a major role by plants in this process. He defined an efficient plant air purifier as:

> *one which could reduce the concentration of a given pollutant from an undesirable level at which harmful effects are observed to a level which would not cause injury or damage, without the air purifier (plant) suffering any ill effects.*

Certain plants can absorb specific air pollutants, that is, hydrogen fluoride, sulfur dioxide, and nitrogen dioxide. The pollutant least absorbed, how-

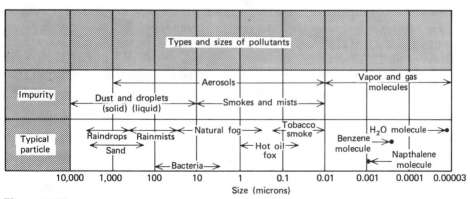

Figure 4.25 Types and sizes of various pollutants encountered in our present-day technological society.

85

Figure 4.26 Proposed schematic for air pollution abatement by plants through the processes of oxygenation and dilution.

ever, is carbon monoxide; and it accounts for roughly one-half the total weight of air pollutants emitted into the air in this country. Species such as sugar maple and yellow birch can readily absorb sulfur dioxide, but they are also easily injured by the gas. Furthermore, it has been estimated that approximately 88 percent of the oxygen produced on the earth through photosynthesis originates in the sea. It has also been estimated that the oxygen produced by an acre of forest land per year represents only 0.03 percent of the total oxygen found over that acre (Weidensaul, 1973). Thus, oxygenation and dilution by plants are probably not at all effective in the abatement of gaseous air pollutants.

Trees, however, are effective in reducing gaseous air pollutants through absorption. A study of ozone pollution and forested areas showed that if an air mass containing 150 ppm of ozone were to stand over a forest for eight hours, the vegetation would absorb approximately 80 percent of it. Taller trees removed more ozone than shorter trees. And the larger and more numerous the leaf stomatal openings, the more effective is the ozone removal (Stevenson, 1970). A recent Russian study has shown that a 1640-ft (500-m) wide green area surrounding factories will reduce sulfur dioxide concentrations by 70 percent and nitric oxide concentrations by 67 percent (Robinette, 1972).

Wind turbulence is a major factor in dispersing gaseous pollutants. Because their presence increases wind turbulence, trees can be used to aid in the dispersal of gaseous pollutants if located downwind from the source of pollution.

Particulate air pollutants can be reduced by the presence of trees and other plants in several ways. They aid in the removal of airborne particulates

86

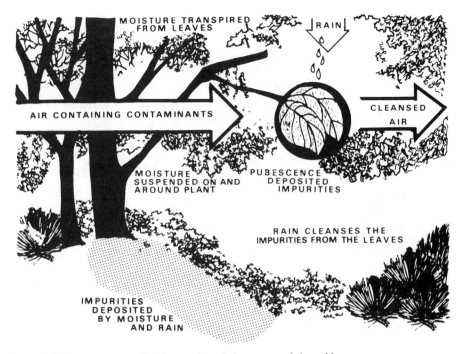

MOISTURE TRANSPIRED FROM LEAVES

RAIN

AIR CONTAINING CONTAMINANTS

CLEANSED AIR

MOISTURE SUSPENDED ON AND AROUND PLANT

PUBESCENCE DEPOSITED IMPURITIES

RAIN CLEANSES THE IMPURITIES FROM THE LEAVES

IMPURITIES DEPOSITED BY MOISTURE AND RAIN

Figure 4.27   Plants reduce air pollutants through the process of air washing.

such as sand, dust, fly ash, pollen, and smoke. Leaves, branches, stems, and their associated surface structures (i.e., pubescence on leaves) tend to trap particles that are later washed off by precipitation. Trees also aid in the removal of airborne particulate matter by air washing. Transpiration increases humidity, thus aiding in the settling out of airborne particulates (Figure 4.27). The results of these processes can be readily observed on trees adjacent to factories or along gravel roads.

Trees also often mask fumes and disagreeable odors by replacing them with more pleasing foliage or floral odors or by actual absorption (Figure 4.28). When trees are planted to aid in air pollution abatement, the following guidelines should be used (Bernatsky, 1968):

1. Plantings should be perpendicular to the prevailing winds.
2. Open and permeable plantings should be combined with dense barrier stands.
3. Plantings should be concentric around the pollution source.

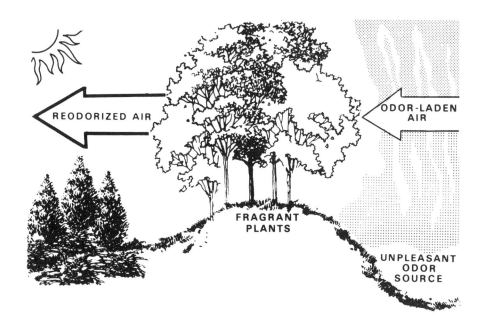

REODORIZATION

Figure 4.28 Trees can reduce unpleasant odors either by absorption of gaseous pollutants or by masking them with their own odors.

Glare and reflection control

Solar radiation affects our visual comfort as well as our thermal comfort. We are surrounded by a myriad of shining surfaces—glass, steel, aluminum, concrete, and water—all capable of reflecting light. We experience discomfort when the sun's rays are reflected toward us by these surfaces. At night, we have to contend with glare from automobile headlights, streetlights, buildings, and advertising signs.

Glare intercepted directly from the light source is termed "primary" (Figure 4.29). Primary glare that is diffused by atmospheric particles makes the light source unrecognizable. This results in a bright haze that can be extremely discomforting.

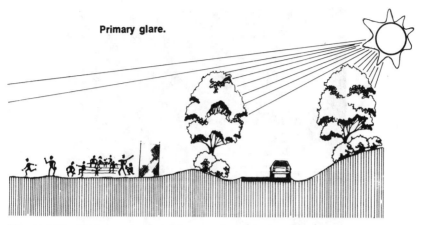

Primary glare.

Figure 4.29 In urban environments trees can be used to intercept primary glare.

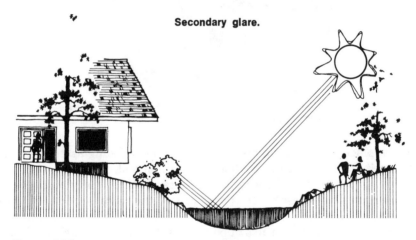

Secondary glare.

Figure 4.30 If positioned correctly, trees can be used to intercept unwanted secondary glare.

Secondary glare is reflected light (Figure 4.30). Factors that affect secondary glare are: (1) smoothness of the surface; (2) angle of the light path; (3) amount of initial light; (4) air temperature; (5) seasonal and diurnal conditions (position of the sun); and (6) atmospheric conditions.

Discomfort from glare and reflection can be reduced architecturally by awnings, window shades, screens, or by proper building orientation and

GLARE
REDUCTION

Figure 4.31 Nighttime glare can be reduced by judicious plant selection and placement.

window placement. We can also wear our polarized sunglasses or purchase tinted windows for our offices, homes, or automobiles.

Plants can be used to screen and soften both primary and secondary glare. Their effectiveness depends primarily on their size and density. Sources of glare must be identified before the proper plants can be selected to control it. The degree of control is also important as it may be desirable to eliminate glare completely or to create a filtering or softening effect.

Plants can block or filter primary glare anytime during the day (Figure 4.29) or night (Figure 4.31). Plants should be selected for proper height and foliage density so that they provide protection throughout their lifetime. Plants can be used along the highway to control early morning and later afternoon glare (Figure 4.32). Night light control can be achieved by the proper placement of trees and shrubs around terraces, patios, windows, or even along streets to protect driver vision.

Secondary glare can be controlled by plantings that intercept the pri-

GLARE CONTROL FOR HIGHWAYS

Figure 4.32 The proper placement of trees along highways and median strips can alleviate early morning and late evening glare hazards.

mary light source before it reaches the reflector or after it strikes the reflector and travels to the viewer's eye (Figure 4.30).

Traffic control

While adding to environmental esthetics, trees and shrubs may be used to aid in traffic control. This includes not only vehicular traffic but also pedestrian and animal traffic (Figure 4.33). Plants enhance the beauty of an area when used to direct people through it in a definite pattern. Other elements often used to perform the same function (wires, fences, and chains) tend to destroy natural beauty.

Before one considers which plants to use in traffic control, the degree of control desired must be ascertained. Can it be accomplished by ground cover, a hedge, or a belt of trees? Once the type of barrier has been selected, species selection should be based on the following considerations (Robinette, 1972):

1. Ultimate height needed.
2. Spacing or planting density needed.

91

TRAFFIC CONTROL

Figure 4.33 Trees and shrubs can be used to direct or channel pedestrian traffic.

3. Eventual desired width of the planting.
4. Characteristic of the plant variety.

Some plants are more suitable for traffic control than others because of such characteristics as branching habit, presence or absence of thorns, and stiff or flexible branches. Robinette (1972) has suggested that plant selections should be based on a 10-point scale for each of the following characteristics: thorniness, mature height, spacing, and mature width. His system is as follows.

Thorn and Branching Habit: Since some plant materials are nearly equal in traffic control efficiency, they should be placed on a 10-point characteristic scale. These 10 divisions are:

1. Single-stem—no thorns—flexible branches.
2. Single-stem—no thorns—stiff branches.
3. Multistem—no thorns—flexible branches—open.
4. Multistem—no thorns—flexible branches—less open.
5. Multistem—no thorns—stiff branches—dense.
    Single-stem—thorns—flexible branches.
6. Single-stem—thorns—stiff branches.
7. Multistem—thorns—flexible branches—open.

92

8. Multistem—thorns—flexible—less open.
Multistem—thorns—stiff branches—open.
9. Multistem—thorns—flexible—dense.
Multistem—thorns—stiff—less open.
10. Multistem—thorns—stiff branches—dense.

Plant Height:
The plants' ultimate height, when mature, is the next factor to be rated. In rating plants on height, the system extends from 3 in. (7.6 cm) or less, to 5 ft (1.5 m) or more.
The 10 dimensions are:

1. 0 to 3 in. (0 to 7.6 cm)
2. 3 to 6 in. (7.6 to 15.2 cm)
3. 6 to 12 in. (15.2 to 30.5 cm)
4. 12 to 18 in. (30.5 to 45.7 cm)
5. 18 to 24 in. (45.7 to 61 cm)
6. 24 to 30 in. (61 to 76.2 cm)
7. 30 in. to 3 ft (76.2 to 91.4 cm)
8. 3 to 4 ft (0.9 to 1.2 m)
9. 4 to 5 ft (1.2 to 1.5 m)
10. 5 ft (1.5 m) and higher

Spacing or Density:
Spacing is important to consider. Plants, regardless of their characteristics, are ineffective in controlling traffic if they are sparsely spaced, allowing movement through the openings between them. In the rating of plants for spacing, the system extends from greater than 30 in. (76.2 cm) to less than 6 in. (15.2 cm).
The ratings are:

1. 30 in. (76.2 cm) and wider
2. 27 to 30 in. (68.6 to 76.2 cm)
3. 24 to 27 in. (61 to 68.6 cm)
4. 21 to 24 in. (53.3 to 61 cm)
5. 18 to 21 in. (45.7 to 53.3 cm)
6. 15 to 18 in. (38.1 to 45.7 cm)
7. 12 to 15 in. (30.5 to 38.1 cm)
8. 9 to 12 in. (22.9 to 30.5 cm)
9. 6 to 9 in. (15.2 to 22.9 cm)
10. 6 in. (15.2 cm) and narrower

Width:
Plant width is also a major factor in determining the effectiveness of plants for controlling traffic. For example, grass or a ground cover may

93

be an effective barrier if the lawn or planting bed is wide enough; yet in a narrow area, even a medium-size hedge may not be adequate.

In the rating of planting widths the system extends from "1" to "10" as follows:

1. 6 in. (15.2 cm)
2. 9 in. (22.9 cm)
3. 12 in. (30.5 cm)
4. 18 in. (45.7 cm)
5. 2 ft (0.6 m)
6. 3 ft (0.9 m)
7. 4 ft (1.2 m)
8. 6 ft (1.8 m)
9. 8 ft (2.4 m)
10. Wider than 8 ft (2.4 m)

Some plants have a spread or width of greater than 8 ft (2.4 m), making them effective barriers when planted in a single row. Others with less width must be planted in multiple rows.

Summary: Since the relative effectiveness of plants in controlling pedestrian traffic can be rated on the basis of a possible 10 points for each of the four factors, 4 is the lowest possible rating and 40 designates a perfect traffic control plant. The total rating system is:

T-C Number (Traffic Control)	Degree of Effectiveness
4–10	Minimum
10–20	Average
20–30	Good
30–40	Excellent

An example of a shrub with an excellent rating would be *Rhamnus carthartica* or buckthorn. It is a large, well-rounded, and fairly dense shrub that has branches and foliage reaching from the ground to a height of 12 ft (3.7 m) and has a 10- to 12-ft (3- to 3.7-m) spread. It yields a characteristic rating of 9, height 10, density 6, width 10, for a combined total of 35.

Some work has been done concerning the use of plants as effective vehicle barriers. One study was conducted using multiflora rose (*Rosa multiflora*) (White, 1953). A series of crash tests were used to determine this species' ability to absorb and decelerate a vehicle with minimum damage. In tests, it performed quite well at speeds up to 47 miles (75.6 km) per hour.

In most instances, roadside plantings should involve resilient plants rather than rigid ones to prevent passenger injury and to minimize automobile damage as much as possible.

Architectural Uses

In building design, materials such as wood, masonry, steel, or concrete are used architecturally as well as structurally. The designing architect asks such questions as: Is privacy needed by the users in a given area? Are there undesirable views that need screening? Is the area too large for comfortable use? Is there a view that can be made more interesting by progressively revealing it? In many situations, trees and shrubs can perform the same architectural functions as do standard building materials.

Each species has its own characteristic form, color, texture, and size. Plants can vary in their use potential as they grow or as the seasons change. Their proper use will vary with the designer and user. Trees, when used in a group, can form canopies or walls of varying texture, height, and density. Some functions can be performed by one tree and others may require many trees. Because they are alive and growing, trees and shrubs are dynamic with regard to their functionality in architectural design.

Since trees and shrubs have architectural potential, they can be used individually or collectively as architectural elements performing the following functions: space articulation (defining space); screening; privacy control; and progressive realization or enticement.

Our perception is based primarily on our function of sight. Space or spatial realization is perceived as the distance to any element that blocks the view (Figure 4.34). Thus, space as perceived in an open park is different from one with many trees. This space realization is further enhanced by textures and shades. Coarse textures and darker shades appear to advance to the viewer, while fine textures and light shades seem to retreat. Trees and shrubs form walls and canopies in landscapes and, along with other architectural components, can be used to enclose, contain, enframe, link, enlarge, reduce, and articulate exterior space. Robinette (1972) discusses these uses along with some examples:

> *Closure is using plants to finish off a space that has been left open. It makes the space more complete and identifiable. Containment is providing a small space within a larger space; thus, attention is toward a small, human-scale feeling of the space [Figure 4.35]. Enframement is a technique used to draw attention to the most important space in the area. It focuses attention on a desired view*

95

Figure 4.34 Trees and shrubs can create a feeling of solitude by their use in breaking up large spaces into smaller ones we are more comfortable with.

Figure 4.35 Trees and shrubs can be used to create unique settings in the process of defining space.

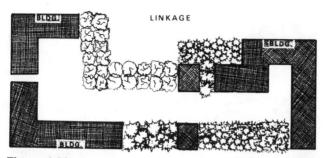

Figure 4.36 Plants can be used to link and unify separate entities.

or object and away from the larger space in which the view or object occurs. Linkage is a technique used to join one space with another to make a large area seem smaller and less alien [Figure 4.36]. Enlargement is a method of changing the apparent size of a large space by contrasting it to an infinite space of the skyscape and making it seem smaller in comparison [Figure 4.37]. Reduction involves the placement of plants in an overly large space to make the space smaller [Figure 4.37]. Plants may be used to subdivide or divide space either horizontally or vertically to reduce the apparent size of the space [Figure 4.38]. In space articulation, plants can also be used to subdivide space three dimensionally. A good example of this is the use of plants to accentuate horizontal space [Figure 4.39]. Plants can also be used to break larger spaces into smaller, irregular or rhythmically distinguishable units. Good examples of this are most park designs where different recreation units are subdivided. Trees and shrubs are often used in home and building settings to articulate or define entrances [Figure 4.40]; in functional realization to provide direction [Figure 4.41]; or channel movement.

Perhaps one of the major uses we normally associate with trees and shrubs is screening. This involves not only screening for view but also for privacy. The selection of plant materials for screening has to be judicious since the view is perhaps objectional for only a part of the year; or it may be one that requires continuous screening. Most are well aware of the highway beautification programs of the 1960s aimed at screening objectionable views from the landscape (junk yards, factories, parking lots, cooling towers for air conditioners, service areas and transformer yards, just to mention a few). Screening visually blocks out an unsightly view and replaces it with a less offensive one (Figure 4.42). It hides the unwanted and allows free

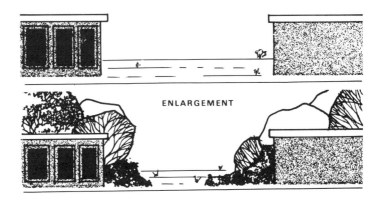

ENLARGEMENT

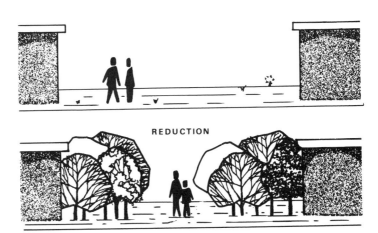

REDUCTION

Figure 4.37 Trees and shrubs can be used to make buildings seem to occupy more space than they actually do, or to reduce open spaces so that they are more pleasant to experience.

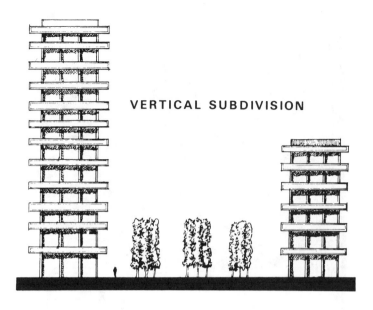

VERTICAL SUBDIVISION

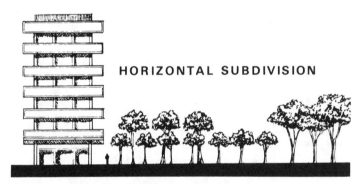

HORIZONTAL SUBDIVISION

Figure 4.38 Trees can be used in inner city areas to break up architectural lines and to unify elements.

Figure 4.39 The proper placement of trees can provide a three-dimensional effect in certain architectural situations.

99

ARTICULATION OF AN ELEMENT
(ENTRANCE)

Figure 4.40 Trees and shrubs can call attention to the location of building entrances.

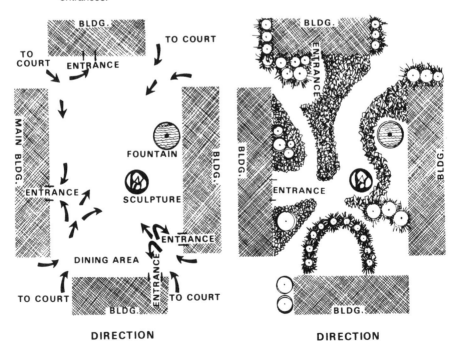

DIRECTION **DIRECTION**

Figure 4.41 Trees and shrubs are especially useful in mall settings to channelize traffic and call attention to entrances.

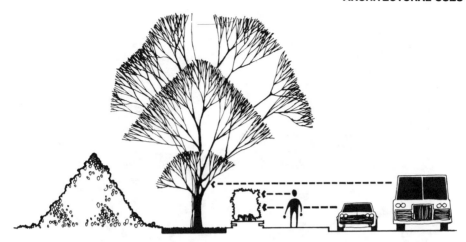

Screen objectionable views.

Figure 4.42 Trees and shrubs are often used to screen undesirable views.

access to the remainder of the surroundings. Screening can be both negative (blocking ugly surroundings from view) or positive (enhancing surroundings).

Privacy control differs from screening in that it secludes a particular area from its surroundings (Figure 4.43). The difference between the two depends on the point of view and intent of either the viewer or the user. As our population continues to increase, there will be a greater need for privacy control as it is essential for survival. It is needed in such activities as sunbathing, camping, picnicking, reading, relaxation, nature watching, and just plain conversation.

In progressive realization, plants are used to direct the unconscious viewer through the landscape, revealing it a little at a time. A view is usually better if enframed or seen through an opening (Figure 4.44). If a landscape view is partially viewed instead of fully revealed, the anticipation enhances

Figure 4.43 Trees and shrubs can be used to provide privacy in urban areas.

101

Figure 4.44 The proper placement of trees can make scenic views even more spectacular because of the window effect they create.

the actual experience. An example is the shadow of a pine branch on a translucent screen. The indirect view is more exciting than a portrait of the branch itself.

Economic Benefits

In a strict economic sense, the net benefits of the urban forest must be those values in excess of their cost of production. Economic benefits could be measured that way, but since urban trees are rarely planted or nurtured solely for economic gain (except in the short run by nurseries and landscape contractors), economics must be kept in context with other values. Thus, a brief discussion will be included here. Monetary values will be more extensively discussed in Chapter 7. Economic benefits can be both direct and indirect. Direct benefits are those derived from the sale or consumption of products such as fuelwood, Christmas trees, or even sawtimber. Indirect benefits occur from savings of energy costs because of shade or wind protection, and from increased property values because of trees. Another indirect benefit is the attraction of business and employees to more esthetically pleasing environments, as has happened in recent years when many corporate offices have moved from older metropolitan areas to more "sylvan" settings in cities such as Atlanta, Georgia and Cincinnati, Ohio.

It should be clearly understood that trees also have negative economic values—generally resulting from the problems they cause for other elements of the urban environment. For example, there are frequent cases of roots in sewers, upheaved sidewalks, limbs interfering with overhead utility wires, branches falling on buildings, cars, and even pedestrians; and there are always leaves to be raked. Such negative values are generally accepted, though, and all benefits, economic and otherwise, are felt to outweigh the cost. If this were not so, there would be few trees in urban areas.

Esthetic Uses

Trees and shrubs provide their own inherent beauty in all settings. They are esthetic elements in our surroundings. They can be beautiful simply because of the lines, forms, colors, and textures they project. Trees and shrubs enframe views, soften architectural lines, enhance and complement architectural elements, unify divergent elements, and introduce a naturalness to otherwise stark settings (Figures 4.45 to 4.48). They also produce unique patterns through reflection from glass and water surfaces and can produce beautiful shadow patterns (Figure 4.49). Trees are also dynamic, giving different appearances in the changing seasons and throughout their span of life. They also provide movement and pleasant sounds—the rustling of leaves and the whistle of wind through a canopy. We are attracted to plants because of many of these characteristics. In addition, plants are useful for the berries, nuts, and shelter they provide to birds and animals (Figure 4.50).

Wildlife Benefits

In his book, *The California Quail*, A.J. Leopold (1977) states that "In my judgment, the creation of living space for a covey of California Quail would represent the gold standard of backyard management." That millions of other Americans share this view through other forms of wildlife is evidenced by the provisions made for songbirds, mammals, and other "wild" creatures in urban areas. Wildlife is abundant in urban areas, some species frequently in higher densities than in rural habitats. Cauley and Schinner (1973) found, for example, racoon densities as high as one animal per two acres in the

Accent and enframe.

Create background setting.

Figure 4.45 Trees and shrubs are used around homes to accent; enframe; or create background settings.

Clifton section of Cincinnati, as compared to densities of one animal per 11.7 acres in natural habitat. In Tucson, Arizona, bird populations (all species) in a 24-block residential area were 26 times higher than in the countryside surrounding the city (Emlen, 1974).

Contrary to the commonly held opinion that urbanization is destructive to wildlife, many species not only adapt to but exploit man-made environments. Urbanization is disruptive, but far from totally destructive to wildlife. Certain species dependent on natural habitats may be lost, but except in extreme cases of permanent removal of vegetation, recolonization, often of other species, occurs. As suggested by the above references to higher than

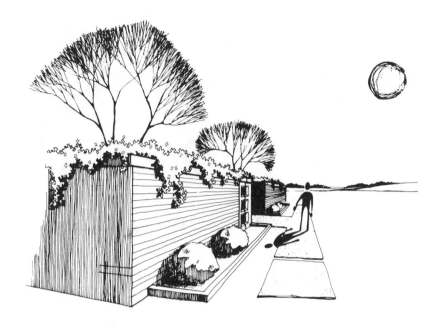

Soften line and mass.

Figure 4.46 Architectural lines can be broken up and their effect softened by proper plant selection and use.

Compliment architecture.

Figure 4.47 Plants can be used to complement modern architectural lines.

105

Unify architectural elements.

Figure 4.48 The use of trees in downtown areas often aids in breaking up straight lines and unifying diverse architectural designs of varying periods.

Create shadow patterns.

Figure 4.49 Not only do trees act as architectural elements but by their presence as shadow patterns, for example, they introduce other elements.

106

Figure 4.50 Plants assist our esthetic enjoyment by enhancing the environment in which our urban wildlife live.

average populations, urban areas are excellent habitats for certain species. Squirrels, racoons, opossums, skunks, reptiles, and a myriad of bird species are common in many cities, depending not only on vegetation, but on man-made structures and human activiites.

People tend to place high values on wildlife in urban areas. In a survey of readers, the *Winnipeg Tribune* found that of the 100 best-loved things in their city of 600,000, "live deer and beaver within the city limits" ranked 48th ahead of the professional hockey team, the former mayor, the new premier, and cheap wine (Shoesmith, 1978).

While esthetic values of wildlife—from the spring songs of birds, to the sight of a mother racoon and her young—are recognized, there can also be problems, as anyone who has dealt with a skunk under the porch or a squirrel in the attic can attest. Far more serious problems can occur too, such as deer causing automobile accidents, birds being sucked into jet engines, and various species of wildlife carrying rabies or other diseases. In Denver, Colorado in 1968, for example, eastern fox squirrels were found to be carriers of rabies (Scott, undated). Until recently, wildlife management in cities concerned itself with such matters of pest control. "But a new concern has emerged for the wildlife biologist: to provide wildlife in urban and suburban areas solely for the enjoyment it provides." This statement and the following by Thomas, DeGraaf, and Mawson (1977) summarize the concept of modern urban wildlife management: Foresters are now looking to trees in cities for more than shade, hardiness, and a minimum of litter. Beyond their visual appeal, trees in urban areas have many amenity values, among them the improvement of air and water quality, modification of microclimate, screening of unsightly land uses, and provision of wildlife habitat.

Wildlife exists as a byproduct of vegetation, and thus the manipulation of vegetation for the above purposes affects the diversity and composition of wildlife populations. Wildlife—especially songbirds—adds color, movement, and sound to the landscape, and thus can contribute much to the human habitability of cities. Basic to the management of desired urban wildlife species is our understanding of their habitat requirements. Such knowledge could be used to manipulate vegetation to provide the amenities of wildlife in addition to those of the vegetation alone.

While vegetation is extremely important to urban wildlife, and its manipulation affords an important opportunity for management, man-made structures are also valuable for certain species: storm sewers and drains for racoons and opossums; chimneys for chimneyswifts; large buildings as nesting places for swallows, sparrows, and even raptors such as owls and falcons. The Peregrine falcons of the Smithsonian Institute's castle in Washington, D.C. are famous examples of the latter. Although positive consideration has been given to wildlife in the design and construction of buildings and other structures, lack of consideration has frequently been disastrous. Television and other communications towers can be especially hazardous to migrating birds. In the fall of 1972, 5,465 dead birds were collected at seven television towers in central Illinois (Bohlen and Seets, 1977). Large windows, and particularly reflective-sided large buildings, can be fatal to birds. Tree reflections in "mirror-sided" buildings have contributed to bird deaths and have, in some cases, resulted in removal of the trees. When reviewed in their entirety, however, man-made structures in urban areas are extremely beneficial to wildlife, particularly when the literally millions of bird houses and feeders and nesting boxes for other animals are considered.

Urban wildlife is a product of the total urban environment and illustrates that no aspect of the urban forest can be considered out of context with others. This is particularly clear in the case of urban fisheries and wetlands management, when water quality, critical to habitat, is influenced strongly by other elements of the urban environment. Water habitat can be extremely important to urban wildlife. Waterfowl exist in surprising numbers in many urban areas. The author counted over 200 overwintering Canada geese in a single flock in Arlington, Virginia during the winter of 1982. Many cities support large populations of ducks, geese, gulls, and shorebirds.

A very important aspect of urban wildlife is the opportunity it presents for education. Franklin (1983) suggested that "fish and wildlife in an urban area can be used as a tool for educating both children and adults—in fact the education of school children can help educate adults. Children participating in urban forest management projects under the supervision of qualified personnel can get their parents interested in conservation affairs."

Urban Wildlife Planning

The topic of urban wildlife planning is well-covered by Leedy, Maestro, and Franklin (1978). We are indebted to them for permission to freely use their material in the following section.

The rapid increase in developed land, particularly in large metropolitan areas, continually decreases the exposure of people to wildlife. While urban areas still occupy a relatively small percentage of the land area in the nation, the percentage of land in such use is growing. Trends suggest that each decade's growth will absorb an area greater than the entire state of New Jersey. This growth would seem to emphasize the need for improving conditions in existing urban areas and for incorporating better planning and environmental management measures in new developments. The provision for wildlife in such areas can enhance the quality of life for people and permit them to better understand and appreciate nature. Historically, the process of urbanization has been carried out with little consideration for fish and wildlife. As a result, populations of some species—all too often the so-called nuisance species—have increased, while other species have declined or, in some cases, been eliminated completely. The effects of urbanization make incorporating fish and wildlife into the planning process all the more important if desirable populations of species are to be maintained for the benefit of people.

Provisions for incorporating wildlife considerations into the planning process for urbanizing areas are based, in large part, on generally accepted wildlife management principles and approaches. There are many approaches to wildlife management. These include the creation of refuges, predator control, artificial stocking, transplanting of wild stock, winter feeding, erection of nesting structures, protection through regulation of hunting, and habitat management. Of these, habitat management is perhaps the most basic, and it is on habitat that the urban or regional planner, landscape architect, developer, and builder have the most impact. It is their designs that determine in large part how and where and to what extent existing habitat will be altered in the course of developing new areas or redeveloping existing urban areas.

There are three levels at which urban wildlife planning takes place: (1) urban or regional planning; (2) site planning; and (3) planning by individual property owners. We will look at the steps involved in the first two and consider some of the tools for their application.

Urban or regional planning is the kind of planning typically carried out by municipal planning departments, county and regional planning commissions, watershed associations, conservation commissions, and state and interstate groups. Of these, it is felt that municipal planning can achieve greater direct implementation of wildlife considerations. Although environmental inventories for planning purposes generally contain statements on wildlife, they are often quite general, and there is often little follow-through to insure the benefits that can be realized by incorporating specific provisions for wildlife in plans and designs. In regional planning, wildlife considerations focus mainly on preservation. A major function of large-scale planning should be to identify key wildlife areas that should be kept free of any type of land development. Major consideration should be given to maintaining a continuous open space/wildlife corridor system throughout the area. There are five general steps in the process:

1. Identify habitats and determine their relative values for wildlife.
2. Identify habitats of threatened and endangered species.
3. Identify groups of plant species of value to wildlife.
4. Analyze adjacent land uses.
5. Develop a continuous open space/wildlife corridor system.

Future developments present planners with the greatest possibility of providing for communities rich in wildlife amenities. It is at the site design level that a more detailed approach to wildlife planning is possible, and consideration can be given to the relationship of infrastructure and other design components to wildlife. While wildlife provisions can be made in any type of site development, it is in residential developments that application is most common. Two basic approaches to residential site design are commonly practiced. One is the conventional subdivision with no provision for open or common space (sometimes called "lot-by-lot" development). The other is cluster, or planned unit development, more commonly referred to as PUD. Projects of this type vary in size from a few acres to the multithousand-acre new town developments.

The PUD planning approach significantly increases the potential for wildlife amenities within the community because of the amount of open space typically provided. It also allows greater opportunities for integrating wildlife because of the flexibility in the design process and general requirements for open space, preservation of natural features, and so forth. Little opportunity for wildlife planning is provided with the conventional subdivision type of development, other than in the landscaping of individual lots. For these reasons, the steps presented as follows apply mainly to PUD type developments, although some application can certainly be made by homeowners.

Wildlife planning must go beyond the identification of important habitats and their incorporation into the open space system. Consideration also must be given to the design of the open space system, the types and locations of all design components (including infrastructure), and proposed management policies and practices. The methodology for wildlife planning at the site level involves some of the same steps as at the regional level. There is need, however, for more detail.

1. Identify habitats on the site and determine their relative value for wildlife. For example: coniferous woodlands (natural or planted); deciduous woodlands by type; mixed coniferous–deciduous woodlands; old fields; meadows; watercourses; impoundments; marshes; wooded swamps; agricultural land.
2. Identify plant species of importance to wildlife as food sources.
3. Analyze adjacent land uses.
4. Identify species on the site and in the region that could be present if a proper habitat was provided.
5. Identify limiting factors for preferred species.
6. Determine how much open space is necessary.
7. Plan the open space system.
8. Integrate wildlife considerations into the conceptual and preliminary designs, giving consideration to minimizing physical impacts and disturbances.
9. Review the architectural design.

If the processes as outlined for both regional and site planning for wildlife are to result in physical implementation, available tools must be used. The most common tools are regulatory mechanisms—zoning and land development ordinances, particularly performance controls built into land development ordinance packages. Ordinances are presently utilized by municipalities for floodplains, steep slopes, woodlands, wetlands, aquifer recharge areas, tree removal (or preservation), sediment and erosion control, open spaces, and so forth. Ordinances of these types impact on wildlife as they influence land use.

If the wildlife planning process is to be effective, there is also the obvious need for communication between urban wildlife specialists, urban planners, designers, and developers. The application is in such infancy that a fundamental need still exists for the instillment in planners of the contribution of wildlife specialists. Conversely, wildlife biologists often do not understand the planning process.*

Other Uses

Although a lengthy list of benefits of the urban forest has been given, there are others. Obviously, there are economic products such as fuelwood, fruits, nuts, Christmas trees, pulpwood, and sawlogs. These will be discussed in detail in Chapter 6. Less tangible, and perhaps no less obvious, are the values of places for children to play, for people to jog or to walk and contemplate nature and their own problems, for lovers to stroll, or for one to be alone (Figure 4.51). Trees are also used as indicators of historic events, as memorials, and as substitutes for the natural environment in inner cities—even on rooftops and balconies (Figure 4.52). Finally, trees also evoke memories of other times, places, and feelings because of the view they present, or a familiar sound, smell, or touch. Who indeed can walk on a leaf-cluttered sidewalk on a rainy autumn day and with the sight and smell of damp leaves not be stirred to remember half-forgotten autumns perhaps long gone by?

Driver and Rosenthal (1978) reported that their studies on the perceived psychological benefits of urban forests and related green spaces indicate that these areas provide a variety of subjectively evaluated benefits, including:

Developing, applying, and testing skills and abilities for a better sense of self-worth.

Exercising to stay physically fit.

Resting, both physically and mentally.

Associating with close friends and other users to develop new friendships and a better sense of place.

Gaining social recognition to enhance self-esteem.

Enhancing a feeling of family kinship or solidarity.

Teaching and leading others, especially to help direct the growth, learning, and development of one's children.

*For a detailed listing of urban wildlife references, see *An Annotated Bibliography on Planning and Management for Urban-Suburban Wildlife,* FWS/OBS–79/25, USDI Fish and Wildlife Service, 1979.

Figure 4.51 All of us have a need for some time to think and reflect on our lives, evoke memories, and commune with nature. Trees and shrubs in urban environments provide appropriate settings.

Reflecting on personal and social values.

Feeling free, independent, and more in control than is possible in a more structured home and work environment.

Growing spiritually.

Applying and developing creative abilities.

Learning more about nature, especially natural processes, man's dependence on them, and how to live in greater harmony with nature.

Exploring and being stimulated, especially as a means of coping with boring, undemanding jobs and to satisfy curiosity and the need for exploration.

113

Figure 4.52 In inner cities, the contrasting settings in which trees are often found express our basic need for the outdoor environment.

Note: Some of the material for this chapter—particularly Figures 4.1 to 4.6 and 4.8 to 4.52—was adapted from *Plants/People/and Environmental Quality,* USDI National Park Service in collaboration with the American Society of Landscape Architects Foundation, 1972.

114

Replenishing adaptive energies and abilities by temporarily escaping adverse social and physical conditions experienced in home, neighborhood, and work environments. Needs include factors such as noise, too many responsibilities, demands of others, time pressures, overcrowding, insufficient green and open space, lack of privacy, pollution, unsafe environment, and demanding jobs.

BIBLIOGRAPHY

Bernatsky, A., "The Importance of Protective Plantings Against Air Pollutants," *Air Pollution Proceedings First European Congress Information Air Pollution. Plants, Animals,* Wagenigen, Wagenigen Center for Agricultural Publications and Documents, the Netherlands, pp. 303–395, 1968.

Bohlen, H.D., and J.W. Seets, "Comparative Mortality of Birds at Television Towers in Central Illinois," *Wilson Bulletin 89,* 1977.

Cauley, Darrel L. and James R. Schinner, "The Cincinnati Raccoons," in *The Metro Forest, A Natural History Special Supplement,* 82 (9): 58–60, 1973.

Cook, David I. and David F. Van Haverbeke, *Trees and Shrubs for Noise Abatement,* University of Nebraska Agricultural Experiment Station Research Bulletin No. 246, 77 pp., 1971.

Cook, David I. and David F. Van Haverbeke, *Tree-Covered Land-Forms for Noise Control,* University of Nebraska Agricultural Experiment Station Research Bulletin No. 263, 47 pp. 1974.

Cook, David I. and David F. Van Haberbeke, *Suburban Noise Control With Plant Materials and Solid Barriers,* University of Nebraska Agricultural Experiment Station Research Bulletin No. EM100, 74 pp., 1977.

Davis, Donald D., "The Role of Trees in Reducing Air Pollution," in *The Role of Trees in the South's Urban Environment,* A Symposium Proceedings, University of Georgia, 1970.

Dewalle, D.R., G.M. Heisler, and R.E. Jacobs, "Forest Home Sites Influence Heating and Cooling Energy," *Journal of Forestry,* 81 (2): 84–85, February 1983.

Driver, B.L., and D. Rosenthal, "Social Benefits of Urban Forests and Related Green Spaces in Cities," in *Proceedings of the National Urban Forestry Conference,* Volume I, ESF Publication 80–003, SUNY, Syracuse, N.Y., 1978.

Embleton, T.F.W., "Sound Propagation in Homogenous, Deciduous, and Evergreen Woods," *Journal of the Acoustical Society of America,* Vol. 33, 1963.

Emlen, J.T., "An Urban Bird Community in Tuscon, Arizona: Derivation, Structure, Regulation," *Condor 76,* 1974.

Eyring, Carl F., "Jungle Acoustics," *Journal of the Acoustical Society of America,* Vol. 18, 1946.

Federer, C.A., "Effects of Trees in Modifying Urban Microclimate," in *Trees and Forests in An Urbanizing Environment,* University of Massachusetts Cooperative Extension Service Monograph No. 17, pp. 23–28, 1971.

Federer, C.A., "Trees Modify the Urban Microclimate," *Journal of Arboriculture,* 2:121–127, 1976.

Franklin, T.M., "Managing Urban Forests for Wildlife," in *Proceedings of Second National Urban Forestry Conference,* American Forestry Association, 1983.

Herrington, Lee P., "Vegetation and Urban Physical Environment," in *Proceedings Urban Forestry Conference* State University of New York, pp. 54–59, 1973.

Jensen, M., *Shelter Effect: Investigations into the Aerodynamics of Shelter and Its Effects on Climate and Crops,* 84 p., Copenhagen, 1954.

Kramer, P.J. and T.T. Kozlowski, *Physiology of Trees,* 642 pp., McGraw-Hill, New York, 1970.

Leedy, D.L., R.M. Maestro, and T.M. Franklin, *Planning for Wildlife in Cities and Suburbs,* USDI Fish and Wildlife Service, FWS/OBS–77/66, 1978.

Leopold, A.S., *The California Quail,* University of California Press, Berkeley, 1977.

Nutter, Wade L., "The Role of Trees in Neutralizing Wastes: A Discussion of Land Disposal Systems for Sewage Effluent and Other Wastewaters," in *The Role of Trees in the South's Urban Environment,* A Symposium Proceedings, University of Georgia, pp. 20–30, 1970.

Owen, Oliver S., *Natural Resource Conservation: An Ecological approach,* 2nd ed., MacMillan, New York, 1975.

Parker, John H., "Landscaping to Reduce the Energy Used in Cooling Buildings," *Journal of Forestry,* 81 (2): 82–83, February 1983.

Scott, J., *Wild in the Streets,* Colorado Division of Wildlife, Denver, Colorado, undated.

Shoesmith, M.W., "Wildlife Management Conflicts in Urban Winnipeg," in *Wildlife and People, Proceedings of the John S. Wright Forestry Conference,* Purdue University, Lafayette, Indiana, pp. 49–50, 1978.

Thomas, J.W., R.M. DeGraaf, and J.C. Mawson, *Determination of Habitat Requirements for Birds in Suburban Areas,* USDA Forest Service Research Paper NE–357, 1977.

5
ENVIRONMENT OF
THE URBAN FOREST

As suggested at the outset of Chapter 2, the urban forest requires a rather complex definition. And the complexity inherent in a definition reflects the complexity of the environment of the urban forest. Brown and Perschel (1982), in attempting to come to grips with this, offered the following:

> *Attempts to define the urban forest quickly indicate that it cannot be defined along one dimension; its inherent complexity warrants an equally complex definition. The urban forest must be defined along a number of continua. For example, urban areas can be measured with regard to their ecological complexity. The degree of complexity may be determined by presence of undergrowth, natural regeneration, percent of crown cover, or suitability for wildlife. In general, the percentage of the area in the more ecologically complex classes increases as one moves along the continuum away from the urban environment.*
>
> *In addition to ecological complexity, other continua should be considered: community infrastructure, human population density, size of parcel, ownership, and social characteristics of associated human populations. It is only by analyzing urban areas in regard to each of these continua that the urban forest can really be understood.*

Although the authors were using the above to support an approach for proper management—and we will deal with this in Chapter 6—they make an excellent statement regarding the physical and human environment of the urban forest. The urban forest exists in an environment dominated by people. Thus, the forest must grow in the spaces left over after all the structures necessary to urban society have been built. The forest must be molded to fit these spaces as it cannot interfere unduly with streets, sidewalks, parking lots, wires, pipes, and other objects. The forest must also withstand compacted soils and polluted air, and it must be compatible with people, pets, and vehicles. The environment of the urban forest, then, involves site, space, and people. We will first discuss the physical environment, that of site and space.

Physical Environment

The construction of cities has dramatic influences on existing and future forest sites (Figure 5.1). Grades are changed, drainage is altered, soil is rearranged and compacted, reflective and absorbable surfaces are created, and air circulation patterns are changed. These influences are most often detrimental to the urban forest. They can, however, be beneficial in that

Figure 5.1   Urban construction has dramatic influences on forest sites.

buildings and other structures sometimes create favorable microclimates for tree growth.

To illustrate these influences, consider the impact of subdivision development in a naturally wooded area: A substantial part of the forest must be physically removed to make room for houses, streets, driveways, patios, pools, water lines, telephone and electricity lines, sanitary and storm sewers, and gas lines. The sudden removal of strips and patches of a natural forest creates edges at which previously shaded trees are exposed to open sunlight and wind. Excavation and grading cause subsoil to replace topsoil and often drastically alter surface and internal soil drainage. Trees to be retained are often damaged by physical contact with machinery, by root exposure or coverage, and by indiscriminate pruning.

The quality of such development into woodlands is as varied as the number of developers and the regulations under which they operate. In most large urban areas one can find excellent examples of subdivisions having been laid gently into the forest. In these same areas, however, some developers have considered the forest a liability to be removed and the land is leveled.

The spatial and site environment can perhaps be best understood by a discussion of the factors that influence the location of trees in the urban

119

forest. These factors apply both to trees left from urban development into native forests and to the selection of trees for planting.

The first requisite of an urban tree is that it must be able to survive and grow under the conditions of its particular site. It must be adaptable to a particular hardiness zone. It should possess a suitable form for its available space, and it should have a root system compatible with soil space restricted because of concrete, asphalt, and underground utilities. The choice of trees must depend on the limiting factors of the site. The fewer the limiting factors, the greater the choices.

Space

Space is probably the most critical factor in locating urban trees. Space is limited physically by buildings, neighboring trees, overhead wires, curbs, sidewalks, and underground utilities. It is also limited by other spaces appropriated for line of sight vision for traffic signals, signs, street lights, and vistas, and by spaces for vehicle and pedestrian clearance. There are two choices: trees can be selected that when mature will fit available spaces; or attempts can be made to make trees conform by pruning them or otherwise controlling their growth. The former is preferable in cases where trees are to be planted. The latter may well be best in areas of new construction where existing trees may be left. However, these choices are often poorly made, and virtually every city has frequent examples of trees located in spaces much too small for their mature growth. The attempts at conforming such trees to available spaces is perhaps the most costly aspect of urban forestry.

For planting purposes, trees can be divided by mature height into three size classes: (1) small trees, less than 30 ft (9.14 m) when mature; (2) medium trees, 30 ft (9.14 m) to 60 ft (18.28 m); and (3) large trees, 60 ft (18.28 m) and over. Such divisions are not rigid as localized sites might cause differences in growth, particularly with "medium" and "large" trees. However, they are convenient groupings to meet spatial requirements. A sample grouping of popular urban species is given below.

Examples of Tree Species
by Size Catagories

Small Trees	
Apricot, flowering	*Prunus armeniaca*
Cherry, "Kwanzan"	*Prunus serrulata*
Crabapple, flowering	*Malus, spp.*
Dogwood, flowering	*Cornus florida*

Hawthorn, Washington	*Crataegus phaenopyrum*
Maple, amur	*Acer ginnala*
Maple, Japanese	*Acer japonicum*
Mimosa	*Albizia julibrissin*
Mountain-ash, European	*Sorbus aucuparia*
Peach, flowering	*Prunus persica*
Pear, bradford	*Pyrus calleryana*
Plum, purpleleaf "Newport"	*Prunus X-blireiana*
Redbud	*Cercis canadensis*
Russian-olive	*Eneagnus angustifolia*

Medium Trees

Ash, green	*Fraxinus pennsylvanica*
Ash, white	*Fraxinus americana*
Baldcypress	*Taxodium distichum*
Birch, paper	*Betula papyrifera*
Catalpa	*Catalpa speciosa*
Chinese chestnut	*Castanea mollissima*
Ginkgo	*Ginkgo biloba*
Hackberry	*Celtis occidentalis*
Honeylocust	*Gleditsia triocanthos*
Kentucky coffeetree	*Gymnocladus dioicus*
Linden, American	*Tilia americana*
Linden, littleleaf	*Tilia cordata*
Locust, black	*Robinia pseudacacia*
Maple, Norway	*Acer platanoides*
Maple, red	*Acer rubrum*
Maple, sugar	*Acer saccharum*
Pagodatree, Japanese	*Sophora japonica*
Poplar, white	*Populus alba*
Tree-of-heaven	*Ailanthus altissima*
Walnut, black	*Juglans nigra*
Willow, weeping	*Salix babylonica*

Large Trees

Cottonwood, eastern	*Populus deltoides*
London planetree	*Platanus x-acerifolia*
Maple, silver	*Acer saccharinum*
Oak, bur	*Quercus macrocarpa*
Oak, English	*Quercus robur*
Oak, pin	*Quercus palustris*
Oak, red	*Quercus borealis*
Oak, white	*Quercus alba*
Pecan	*Carya illinoensis*
Sweetgum	*Liquidambar styraciflua*
Sycamore	*Platanus occidentalis*
Tulip poplar	*Liriodendron tulipifera*

Mature height, although extremely important, cannot be the only factor in fitting trees to space. Form must also be considered.

Tree form is an important element in landscape design and should be given careful consideration when fitting trees to available spaces. Trees assume seven basic forms: (1) irregular; (2) vase; (3) oval; (4) pyramidal; (5) columnar; (6) round; and (7) weeping. Many species can be identified by their characteristic form, although varietal differences are sometimes confusing. Form can also be impaired by crowding, storm breakage, or other influences. Recently, much work has been done by nursery workers and others in selecting and breeding trees for specific forms. Fastigate Washington hawthorn, ascending Norway maple, columnar red and sugar maples, and columnar English oak are some popular results of this work.

For confined spaces such as narrow parkways along streets, trees with columnar, oval, and vase forms are best. Columnar trees provide relatively little shade but are often planted close to each other for screening purposes. They are also used to soften or accent lines of tall buildings. Vase-shaped trees have long been popular for street planting as their ascending branches can be allowed to spread above streetside objects and spaces needed for vehicle and pedestrian passage. This quality helped make the American elm such a popular streetside tree prior to spread of the Dutch elm disease.

Pyramidal trees provide rather poor shade as they are narrow at the top. They are best used as specimen trees in the landscape where adequate room is available for low lateral branch development. Pin oak, sweetgum, and baldcypress are examples of pyramidal trees. They are generally not well-suited for streetside planting on other than wide treelawns. Pin oaks are commonly misplanted along streets with the later necessity of removing lower, drooping branches for traffic clearance. (Size, form, and other design elements are discussed in further detail on pages 150–160.)

Soils

Urban forest soils are often drastically altered by construction activity. Topsoil is often removed entirely or covered with subsoil. Sand, chemicals, and waste building materials are frquently incorporated. Many building sites are in fact "filled" ground from rubble and debris. Severe compaction is common and internal drainage is often restricted. Water tables are also often changed.

Engineering specifications for streets, parking lots, and other areas often require intensive soil compaction. Unfortunately, adjacent areas where trees may be planted are frequently also compacted. Compaction also occurs on building sites simply by the presence of heavy construction machinery.

Tightly compacted soils are poorly aerated, and gaseous exchange between soil air and the atmosphere is restricted. This results in an unfavorable oxygen–carbon dioxide balance that can inhibit tree root growth. Such soils are also poorly drained internally, with water occupying air spaces and further restricting gaseous exchange. Poor soil aeration can also result from a rise in ground water tables. Water tables are frequently altered by construction activity.

Inadequate soil aeration affects trees indirectly too. Soil ecosystems are altered and favorable soil organisms may be destroyed. Certain damaging fungi thrive in poorly aerated soils because natural control conditions are not present. Root permeability may also be impaired because of excessive carbon dioxide.

Special soil treatments and amendments are often needed to assure plant growth. In extreme situations major treatments are necessary, involving extra excavation of planting sites, construction of drainage systems, and replacement with new soil. Soils can be physically improved by incorporation of sand, peat, or other materials. Organic matter is often needed, but adding it will have little benefit if drainage, fertility, and acidity are not also corrected. Peat moss, stable manure, and compost are commonly used organic amendments. Mineral deficiencies can be corrected by the addition of commercial fertilizers. Soil acidity must also often be altered as it largely governs the availability of nutrients.

Topography

Topography exerts a strong influence on the urban forest, particularly at its extremes of lowlands subject to flooding, steep hillsides prone to erosion, and high elevations where climatic influences are more critical. It is hardly enough to recognize that portions of the urban forest are subject to flooding or that landslides on steep slopes do occur. The urban forest manager must understand the cause-and-effect relationships and the often complex limits of control. The complexities of topographical situations and their relationships to urban forestry are illustrated by the Cincinnati Hillside Project. The Cincinnati urban area has about 25 square miles of steep, highly erodable hillsides containing a wide variety of land uses and ownerships. In an effort to ultimately manage these hillsides at the optimum, the Cincinnati Urban Forestry Board, Hillside Trust, and Planning Commission in cooperation with the Northeastern Forest Experiment Station of the USDA Forest Service instituted an intensive study. The study involved (Sanders, 1982): (1) a detailed reconnaissance of urban forest coverage, urban land uses, and specific environmental characteristics on or near the 23 hillsides of Cincinnati; (2) ranking of the hillsides in terms of their relative importance in regulating local air quality, temperature, and wind regimes, including effects on urban energy consumption; in influencing noise levels and property values; and in impacting soil erosion; (3) development of an urban forestry management guide for the hillsides, stating goals, objectives, and choices for each of the hillsides; (4) evaluation of the effectiveness of current local hillside conservation legislation and regulations to determine whether amendments or new legislation were called for; and (5) initiation of a plausible public awareness program. While such studies might well be applicable to other segments of the urban forest, it is because of the special nature of the hillside topography that they were initiated in Cincinnati.

Microclimate

The urban forest exists in a microclimate of itself and human structures. The microclimatic factors having the greatest influence on tree growth are air temperature, humidity, and wind. Microclimates can be either beneficial

Figure 5.2 Street-level microclimate No. 1: areas with extensive evaporative or transpiring surfaces, such as this park in New York City.

or harmful depending on how much the extremes and duration of heat, cold, wind, and evapotranspiration are influenced.

In general, cities tend to be warmer in both summer and winter than the surrounding countrysides. Wind velocity is less and relative humidity is generally lower. Such generalities can be misleading, however, as cities are not single microclimates. Urban areas are agglomerations of man-made structures, land forms, trees, and other organisms. Each location within an urban setting has its own microclimate, depending on the character and arrangement of the various elements. C.A. Federer (1971) has identified three broad classes of street-level microclimates: (1) areas with extensive evaporative or transpiring surfaces—parks, wide streets with trees, and the vicinities of rivers or lakes (Figure 5.2); (2) wide treeless streets, squares, and parking lots, open to the sky and very dry (Figure 5.3); and (3) narrow streets and courtyards surrounded by relatively tall buildings (Figure 5.4).

Because of extensive transpiration, Microclimate No. 1 tends to have lower temperatures and higher humidity in the summer. It is also cooler in winter than other areas of the city. Wind velocities are higher because there are fewer physical barriers. These areas more nearly represent the climate

Figure 5.3 Street-level microclimate No. 2: areas open to the sky and very dry such as this parking lot.

125

Figure 5.4 Street-level microclimate No. 3: narrow streets and courtyards surrounded by tall buildings.

of surrounding countrysides. From a climate standpoint, there would be little advantage or disadvantage for tree growth in these situations.

Microclimate No. 2 has much lower humidity and higher temperatures than other areas. High temperatures tend to be extreme because of excessive radiation. Wind velocities are average to slightly lower than the surrounding countryside. Tree selection is limited here as succulent species often cannot be used.

Microclimate No. 3 has cooler summer temperatures and much greater wind protection than other areas. Winter temperatures are somewhat warmer because of radiated heat from surrounding buildings. A greater variety of tree species can be grown here because of moderated extremes of temperature and wind velocity.

However, there are situations in each area that defy the average. Structures that provide permanent shade, reflected light, or heat radiation create their own mini-microclimates. Also tall buildings and streets often cause a tunneling effect that greatly increases wind velocity. Street direction can have a microclimatic effect, as north-south streets between tall buildings are exposed to midday sun and east-west streets tend to be more shaded.

Herrington (1978) related microclimate to human comfort:

There are at least two different kinds of activity taking place in various urban spaces—amenity behavior where people have made the choice to use a given space for relaxation or recreation and non-amenity behavior where people have to venture into the street environment for more utilitarian purposes. For people engaged in amenity behavior, the design emphasis is on comfort; for non-amenity behavior the emphasis is on creating a shelter from extreme or stress-inducing climatic conditions.

We can combine microclimatic variables into several groups that can define different kinds of "comfort": WIND. The mechanical effects of wind—that is, the force exerted by the wind on a person and the amount of debris in the air; GLARE. The glare and brightness of street and amenity spaces—primarily important in terms of ocular comfort; PRECIPITATION. The degree to which people are pelted by rain; and THERMAL. The thermal comfort of people— related to the sensations of temperature and comfort.

He concludes by saying that urban forest "vegetation can play an important role in providing more comfortable places for people to relax, recreate or travel. The science is mostly complete, and we need to get on with the job of creating design aids and guides."

A comprehensive study was initiated in Dayton, Ohio in 1979 to determine the nature of the city's urban forest resources and how these resources affect the city's natural environment and the overall quality of life. Among the findings and conclusions of this study were the following (Rowntree and Sanders, 1983):

The City of Dayton has a tree crown cover of 22 percent, which works out to a 37 percent "tree stocking level" when artificial surfaces are removed from computation. There is a wide variation in these levels between neighborhoods and within land uses. Canopy coverage varies widely but predictably within land uses. Single family residential and open or vacant spaces have tree stocking levels of approximately 50 percent, all other uses being significantly lower.

Existing tree canopies lower the city's temperature by about 20 percent from what it would be without trees. Modest increases in canopy cover could raise this figure to about 26 percent (about 2° F). In some neighborhoods these figures are approximately doubled. Informed and effective urban forest management could appreciably lower the heat island peak near the center of the city and shrink its overall physical extent.

Pollution

The urban forest is subjected to varying degrees of pollution. Pollutants can be physical, chemical, and physiological and may be defined simply as foreign substances or influences that adversely affect the growth function of plants. Pollution becomes a matter of concern when tolerable limits of adversity are exceeded. Physical pollutants are those that physically impair plant function, such as heavy dust covering leaf stomata and restricting respiration. Chemical pollutants are chemical in origin and primarily chemical in effect. They may cause physical damage in some cases, but their principal impact is on the chemical processes within plants.

Air pollution

Air pollutants that damage trees are primarily gaseous chemicals, although particulate matter is also often involved. Transportation, electricity generation, and industry are the primary sources of phytotoxic air pollutants. The main air pollutants of the urban forest are (1) sulfur dioxide; (2) ozone; (3) fluorides; (4) ethylene; (5) oxides of nitrogen; (6) ammonia; (7) chlorine and hydrogen chloride; (8) particulates; and (9) herbicides. Of these, sulfur dioxide, ozone, and herbicides are of major importance.

Sulfur dioxide comes mainly from the burning of coal for the generation of electricity and from industries that refine petroleum or metals. Trees and other plants appear to be most sensitive to SO_2 in early summer when leaves are expanding. Sulfur dioxide enters leaves through stomata and reacts with cells, causing injury or death to tissues. Injury may be chronic or severe, depending on pollution levels and the resistance of a particular plant. Ash, aspen, birch, American elm, poplars, and white pine appear sensitive to SO_2, while dogwood, juniper, maple, spruce, and sycamore appear relatively tolerant (Table 5.1).

Ozone is a natural and necessary component of the upper atmosphere. At ground level, however, it can cause injury and death to plants. Ozone is formed by photochemical reactions in sunlight from hydrocarbons and oxides of nitrogen that are emitted primarily by automobiles and other vehicles. Ozone has caused widespread injury and death to ponderosa pine near Los Angeles and is related to the temperature inversion—smog phenomenon of that area. Ash, honeylocust, white oak, white pine, sweetgum, sycamore and tuliptree are sensitive to ozone. Black walnut, gray dogwood, balsam and white fir, maple, red oak, and spruce appear relatively tolerant (Table 5.2).

Airborne herbicides are major sources of plant damage to the urban forest. Used for the control of unwanted vegetation, herbicides such as 2,4-

TABLE 5.1

Relative Sensitivity of
Selected Forest Species to SO_2
(Loomis and Padgett, 1975)

Sensitive	Tolerant
Ash	Black tupelo
Aspen	Boxelder
Birch	Dogwood
Blackberry	Juniper
Carelessweed	Maple
Catalpa	Oak, live
Dewberry	Sourwood
Elm, American	Spruce
Larch	Sycamore
Oak, blackjack	Yellow-poplar[a]
Pine, eastern white	
Pine, jack	
Pine, loblolly (seedlings to 6 ft)	
Pine, Virginia (seedlings to 6 ft)	
Poplar	
Ragweed	

[a]Sensitive spring and early summer.

TABLE 5.2

Relative Sensitivity of
Selected Forest Species to Ozone
(Loomis and Padgett, 1975)

Sensitive	Tolerant
Ash	Birch, European white
Honeylocust	Black walnut
Larch, European	Dogwood, gray
Oak, white	Fir, balsam
Pine, Virginia	Fir, white
Pine, eastern white	Maple
Pine, jack	Oak, red
Poplar	Spruce
Sweetgum	
Sycamore	
Yellow-poplar	

D and 2,4,5-T are often volatilized or particles are carried by the wind. Contamination by these chemicals can be particularly troublesome near highways and other public areas where chemicals are used as an economic alternative to mowing or other weed control practices. The urban forests of agricultural regions are often subjected to herbicide drift from aerial or ground application to crop and pasture lands. Redbud and goldenraintree are examples of species that are extremely sensitive to airborne herbicides.

An excellent summary of air pollution and trees (excluding herbicide pollution), including a listing of 43 references, is *Air Pollution and Trees in the East* (Loomis and Padgett, 1975). Much of the previous discussion was based on this publication.

Soil pollution

Soil pollution usually applies to restricted areas and often causes severe problems to trees and other plants. Common soil pollutants are oil, gasoline, manufactured and natural gas, salt, and herbicides. Some pollutants are toxic to plant tissue and cause direct damage. Some are taken systemically through the root system. And others cause indirect damage by displacing oxygen or preventing natural chemical exchanges in the soil.

Soil pollution by herbicides has increased in recent years. A variety of preemergent and postemergent grass and weed killers are now on the market. Although the Environmental Protection Agency requires detailed labeling of these chemicals, the labels (including specific directions for use) are often not read, and damage to nontarget plants results. Occasionally even proper use can result in damage, when heavy rains following applications concentrate herbicides in low-lying areas within the root zones of trees or other plants. Soil sterilants, used under fences, around signs, parking lot posts, and other areas are often translocated or leached into the root zones of trees. Damage is by systemic intake through the roots.

Rock salt has long been used for melting snow and ice from vehicular and pedestrian ways. Plant damage can be caused by the runoff of salty water and salt absorption via the roots. Damage can also occur from the salt sprayed by passing vehicles and is thus a localized form of air pollution (Lumis, Hofstra, and Hall, 1975). The relative salt tolerance of trees is given in Table 5.3.

TABLE 5.3
Relative Salt Tolerance of
Some Selected Trees
(Based on Dirr, 1976)

Species	Salt-Tolerance Rating[a]		
	Good	Moderate	Poor
Ailanthus	*		
Alder, European black			*
Ash, American	*	*	
Ash, green	*	*	
Basswood, American		*	*
Beech, American		*	*
Birch, European white	*	*	
Birch, paper	*	*	
Birch, sweet	*		
Birch, yellow	*		
Buckthorn	*		
Catalpa, northern		*	
Cherry, black	*		*
Crabapples		*	*
Douglasfir		*	
Elm, American		*	*
Elm, Siberian	*	*	
Fir, balsam		*	*
Hackberry, common			*
Hawthorns			*
Hemlock			*
Hickory, shagbark	*		
Holly, American	*		
Honeylocust	*		
Hornbeam, American			*
Horsechestnut	*		
Larches	*		

TABLE 5.3 (continued)

Species	Salt-Tolerance Rating[a]		
	Good	Moderate	Poor
Locust, black	*		
Magnolia, southern	*		
Maple, amur			*
Maple (boxelder)		*	*
Maple, Norway	*	*	
Maple, red		*	*
Maple, silver	*	*	*
Maple, sugar	*		*
Metasequoia			*
Mountain ash		*	
Mulberry, white	*		*
Oak, blackjack	*		
Oak, bur	*	*	*
Oak, chinquapin			*
Oak, English	*		*
Oak, northern red	*		
Oak, pin			*
Oak, swamp white			*
Pears		*	
Pine, Austrian	*		
Pine, eastern white			*
Pine, jack	*		
Pine, mugo	*		
Pine, pitch	*		
Pine, ponderosa		*	
Pine, red			*
Pine, Scotch	*	*	*
Planetree, London			*
Poplars	*	*	
Plum, purpleleaf	*		
Redbud, eastern			*
Redcedar, eastern	*	*	
Serviceberry	*		
Spruce, blue	*	*	
Spruce, Norway		*	*
Spruce, white		*	*
Sumac, staghorn	*		
Tupelo, black	*		
Walnut, black	*		*
Walnut, English	*		*
Willows	*	*	*
Yellow-poplar			*
Yew		*	*

[a]Species with dual entries indicate different findings by various authors.

132

Manufactured gas (primarily from coal) contains hydrogen cyanide, carbon monoxide, and unsaturated hydrocarbons. When leaked into the soil, these chemicals can be toxic to plant roots. When used extensively for illumination in the first third of this century, manufactured gas was a significant soil pollutant. Restricted now mainly to containerized use in rural areas, manufactured gas is a minor cause of plant injury. It may, however, cause significant plant injury if it emerges again as a major source of heat energy when natural gas reserves are depleted and the use of coal intensifies.

Natural gas is not believed to be toxic to plant roots (Pirone, 1972). However, it is believed that damage can occur in the vicinity of natural gas leaks from excessive soil drying or by displacement of oxygen. The damages caused by manufactured and natural gas, however, are still confused.

Light pollution

Illumination of urban areas for reasons of comfort and security has increased dramatically in recent years. Since 1950, mercury, metal halide, and sodium lamps have come into common use and have altered the environment of portions of the urban forest. Of particular concern are plants subjected to 24-hour light. In the sense that such conditions may be damaging to plants, light must be considered as pollution. Plant reaction to light is in response to a combination of quality, intensity, and duration interacting with the environment (Cathey and Campbell, 1975). Any factor that would limit growth, such as heat, cold, and drought, could override the effects of light. Plants react differently, but continuous lighting, when other growth conditions are favorable, tends to promote lengthening of branch internodes and expansion of leaf area. Formation and maintenance of chlorophyll are also depressed. The danger of such growth responses is that plants may be more sensitive to air pollution or that growth may be vulnerable to early fall freezes. Amur maple, white and paper birch, sycamore, American and Siberian elms, and flowering dogwood are common urban trees that are sensitive to artificial light (Table 5.4).

People

At the beginning of this chapter, it was stated that the urban forest exists in an environment dominated by people and that the forest must grow in spaces that remain after all structures necessary to urban society have been built. The "people factor" of the urban forest environment is largely a product of the values or meanings that people attach to trees. That different people view trees differently was stated by Appleyard (1978): "On first thought you might assume that everyone loves trees and that your own image of trees

TABLE 5.4

Sensitivity of 40 Plants
to Security Lighting
(Cathey and Campbell, 1975)

High
Acer ginnala, amur maple
Acer platanoides, Norway maple
Betula papyrifera, paper birch
Betula pendula, European white birch
Betula populifolia, white birch
Catalpa bignonioides, catalpa
Cornus alba, Tatarian dogwood
Cornus florida, dogwood
Cornus stolonifera, red-osier dogwood
Platanus acerifolia, London planetree
Ulmus americana, American elm
Ulmus pumila, Siberian elm
Zelkova serrata, zelkova

Intermediate
Acer rubrum, red maple
Acer palmatum, Japanese maple
Cercis canadensis, eastern redbud
Cornus controversa, giant dogwood
Cornus sanquinea, bloodtwig dogwood
Gleditsia triocanthos, honeylocust
Halesia carolina, silverbell
Koelreuteria paniculata, goldenraintree
Ostrya virginiana, eastern hophornbeam
Phellodendron amurense, cork tree
Sophora japonica, Japanese pagodatree

Low
Tilia cordata, littleleaf linden
Carpinus japonica, hornbeam
Fagus sylvatica, European beech
Ginkgo biloba, ginkgo
Ilex opaca, American holly
Liquidamber styraciflua, sweetgum
Magnolia grandiflora, southern magnolia
Malus boccata, Siberian crabapple
Malus sargenti, Sargent's crabapple
Pinus nigra, Austrian pine
Pyrus calleryana, Bradford pear
Quercus palustris, pin oak
Quercus phellos, willow oak
Quercus robur, English oak
Quercus shumardi, shumard oak
Tilia europaea, European linden

Note: Plants have been listed alphabetically and are not grouped in descending order of sensitivity. A high, intermediate, or low rating identifies the relative responsiveness of the plants to security lighting. Plants with low sensitivity are preferred in areas with security lighting.

is shared by all others. Though trees are among the most popular elements of the urban environment, quite a few citizens find them to be a problem." He goes on to list the many meanings of trees to people:

Trees as natural forms and masses.

Trees as shaped ornaments.

Trees as incense and music.

Trees to mask ugliness.

Trees as a refreshing contrast.

Trees as a threat to urbanism.

Trees as shade, shelter, and protection.

Trees as hazard.

Trees as polluters.

Trees as obstacles and intruders.

Trees as producers of fruits, nuts, and leaves.

Trees as wood.

Trees to climb and hide in.

Trees to be planted, cultivated, and managed.

Trees as symbols of innocence in nature.

Trees as symbols of self and others.

Trees as alien symbols.

Trees as a social token.

Trees as obscure Latin names.

Trees as education.

Trees as jobs and economic value.

Trees with no meaning.

These meanings affect the urban forest profoundly, as any one of them, if shared by the right people at the right time, can influence what is done to the forest. Hence, the manager of the urban forest must be aware of them, for all of what is done must reflect the values of people.

Fortunately for the urban forest, the majority of people value it positively, and although their activities are often negative factors, they are more than offset by the good accomplished. Thus, people dig, push, and pollute, but they also plant, prune, fertilize, and do a half dozen other necessary

135

things to this forest. If human beings and the urban forest are to be compatible, the forest must be managed. The management of the urban forest is considered in Chapter 6.

BIBLIOGRAPHY

Appleyard, D., "Urban Trees, Urban Forests: What Do They Mean?", *Proceedings of the National Urban Forestry Conference,* Volume I, ESF Publication 80–003, SUNY, Syracuse, N.Y., pp. 138–155, 1978.

Brown, C.N., and R.T. Pershall, "Urban Forestry From A Forest Management Perspective," USDA Forest Service, Cooperative Forestry files, Washington, D.C., unpublished.

Cathey, H.M. and L.E. Campbell, "Security Lighting and Its Impact on the Landscape," *Journal of Arboriculture* 1(10): 184, 1975.

Federer, C.A., "Effects of Trees in Modifying Urban Microclimate," *Proceedings of Symposium on the Role of Trees in the South's Urban Environment,* USDA Forest Service, p. 26, 1971.

Herrington, L.P., "Urban Vegetation and Microclimate," *Proceedings of the National Urban Forestry Conference,* Volume I, ESF Publication 80–003, SUNY, Syracuse, N.Y., pp. 256–257, 1978.

Loomis, Robert C. and William H. Padgett, *Air Pollution and Trees in the East,* USDA Forest Service, 1975.

Lumis, M.P., H. Hofstra, and R. Hall, "Salt Damage to Roadside Plants," *Journal of Arboriculture,* 1(1): 14, 1975.

Pirone, P. P., *Tree Maintenance,* Fourth Edition, Oxford University Press, New York, p. 204, 1972.

Rowntree, R.A., and R.A. Sanders, "Executive Summary, Dayton (Ohio) Climate Project," in-service document, USDA Forest Service, NE Forest Experiment Station, SUNY College of Environmental Science and Forestry, Syracuse, N.Y., 1983.

Sanders, R.A., "Master Study Plan—Cincinnati Hillsides Project," in-service document, USDA Forest Service, Northeastern Forest Experiment Station, Syracuse, N.Y., 1982.

6
MANAGEMEMENT OF
THE URBAN FOREST

This chapter is concerned with the management needs of the urban forest and how these needs are met. There are two forms of management: that which is done *for* the forest to maintain health and vigor; and that which is done *to* the forest to prevent undue interference with the trappings of society. This distinction is significant as indeed much of what is done for the forest is directed toward its betterment. However, a great deal of what is done in the name of management is for other reasons. This is perhaps best illustrated by the percentages of municipal forestry budgets allocated to utility line clearance. Often these percentages far exceed the amount of planting, insect and disease control, and all other practices combined.

As discussed in Chapter 2, the urban forest of even a small city or town might have several thousand owners. All owners, from the smallest private landholders to the largest agency of government, have their own purposes of ownership. And these purposes have a profound influence on the management of the urban forest. Americans have long been staunch defenders of individual property rights, including the right to hold land for whatever purpose they desire. Obviously these rights cannot be absolute as individuals also have a responsibility to society. In the case of land ownership, this responsibility is manifested by statutes or ordinances that either prohibit or require certain things for the benefit of society. For example, there are laws prohibiting debris burning within metropolitan areas and laws requiring property owners to pay taxes to provide a variety of services for society.

It is the responsibility of owners of the urban forest to manage their portions of the forest to the minimum extent acceptable to society. It is the right of property owners to exceed this minimum. The standards vary with time and place. For example, townspeople in one community might tolerate dead trees or poison ivy on a vacant lot whereas in another they would not. They might also have tolerated such a situation 20 years ago and today they might not. The reasons are as complex as the factors that influence human values.

In a typical city the urban forest is managed by a myriad of public and private owners with city government having certain authority over all ownerships. This authority stems generally from police powers for the protection of public health, safety, morals, convenience, and welfare. It allows for the protection of the forest from fire, insects, and disease; it requires control of vegetative growth for public safety; and it provides for certain other areas of management. This authority is usually exercised to the minimum possible extent but offers the only direct opportunity for management of the urban forest as a whole. Thus, management of the urban forest is primarily the responsibility of individual property owners with limited involvement by city government except on city-owned lands. The result is a hodgepodge of species, ages, sizes, and conditions varying from lot to lot, neighborhood to neighborhood, to the edge of the city and beyond. On a whole, it is a situation of contrasts, comparisons, and complete differences, every bit as complex as a natural forest.

The urban forest has four fundamental needs: planting, maintenance, protection, and removal. We will consider as management the methods by which these needs are met. We will not attempt to discuss in detail the cultural aspects of these needs as there are many excellent publications on these subjects (please see the list of selected readings at the end of this chapter).* Also we have included in the Appendix the model specifications and standards of arboricultural practice that were prepared for the Atlanta, Georgia area (Sandfort and Nobles, 1976). These are presented as an example of the traditional cultural products of management.

We will consider the alternatives for dealing with the needs of the urban forest, largely from the standpoint of municipal governments—their exercise of responsibility to streetside, park, and other public segments of the forest; and their influence on the privately owned forest. We do so with the clear realization that all other units of government—county, state, and federal— may also be responsible for segments of the urban forest. However, it is municipal government that has the most pervasive influence, by statutory authority and by the activities of a centrally located forestry department.

Urban forest management, as exercised by the municipal government, has as its foundation: ordinances, policies, and traditions (and traditions are usually translated into policies). These are products of the "people factor" of the urban forest environment as discussed in the previous chapter and are parts of the "bag" of tools of management. Management tools are but one part, however, as urban forestry programs must operate on the basis of three elements: (1) resource characteristics; (2) management tools; and (3) the operating environment.

The characteristics of each segment of the urban forest to be managed must be known: location, composition, condition, ownership, uses, values, and external factors such as pending land use changes. Most large municipalities attempt to manage certain segments of the urban forest directly, usually streetsides and parks, with their own crews or by contract; and attempt to influence management of other segments, both public and private, by indirect means, ranging from newsletters to homeowners, to certification and training of commercial arborists, to encouraging local nurseries to stock "new" species.

*A detailed list of further references is given in *Urban Forestry: A Bibliography* by J. Albrecht and P.J. Weicherding, Miscellaneous Publication No. 1, Agricultural Experiment Station, University of Minnesota, 1980, and a *Supplement to Urban Forestry: A Bibliography*, Miscellaneous Publication No. 16, 1982, by the same authors and publisher.

Management tools include ordinances, policies, labor, equipment, contracts, and communications. Since communications are at the heart of indirect management, they warrant further listing and include personal contacts, direct mail, media, workshops, demonstrations, training, and all other ethical ways of influencing human behavior.

The operating environment also involves ordinances and policies, but includes such factors as city administration, advisory boards, citizen's groups, public attitudes, local issues, and perhaps a host of others. It should be the goal of the urban forest manager to improve the operating environment—as a result of the good works of his/her office and as a planned effort (we will treat this further in Chapter 9).

Before dealing with the application of management to meet the cultural needs of the urban forest, let us look at the approaches to segmenting the forest and learning its characteristics. Fundamental to this, however, is an appreciation for the urban forest as a whole and the knowledge that it can be managed in its entirety: directly in part and indirectly in part. It is in this area that the principles and practices of traditional forestry can be best applied—to the benefit of the urban forest, its owners and users, and forestry.

Classifying and Inventorying the Urban Forest

As stated, a thorough knowledge of the urban forest is fundamental to effective management. There are several approaches to dividing the urban forest into segments or management units: by ownership, land use, location, and vegetative or other characteristics. Of these, ownership and land use are probably the most important initial criteria. Further division by vegetation type or condition within ownership or land use segments is often necessary for intensive management. Brown and Perschel (1982) suggest that only by considering the ecological complexity with the factors of community infrastructure, human population density, size of parcel, ownership, and social characteristics can the urban forest be understood, and appropriate management plans for individual parts of the forest be developed. They suggest the following classes of the urban forest: street tree area; yard, courtyard; open park (town green, cemetery, golf course); highway (median strip, margin); wasteland (vacant lot, industrial area, rail right-of-way); forested woodlot (park, estate, arboretum); and forest preserve. Each class is then considered by size, ownership, location relative to human population density, social characteristics of human population, and possible management practices. Examples of three classes are as follows:

Street Tree Area
   Size: one block, continuous sidewalk strip.
   Ownership: municipal.

Location relative to human population density: high density residential.
Social characteristics of human population: low to middle income.
Possible management practices: individual tree maintenance.

Yard, Courtyard
Size: ½ acre.
Ownership: private.
Location relative to human population density: single family houses.
Social characteristics of human population: middle income.
Possible management practices: individual tree management to provide shade and wildlife habitat.

Forested Woodlot
Size: 10+ acres.
Ownership: county.
Location relative to human population density: between subdivisions and towns.
Social characteristics of human population: middle to upper income.
Possible management practices: possible fuel-wood thinnings.

This is an excellent approach and needs only to be followed by a plan for affecting or expediting the management practices identified.

Urban Forest Inventories

Physical inventories of trees and other vegetation on individual units of the urban forest are important to management—the importance being somewhat proportional to intensity of management. The type of inventory depends on the unit class (as discussed above) and the objectives and intensity of management. In some situations, such as shown for the forested woodlot class above, traditional timber cruises may be appropriate.* In others, such as perhaps highway medians with few trees, informal tree counts might be used. In units of intensive direct management, however, such as the above street tree area class, detailed single-tree inventories are needed. It is to this urban forest class that the following discussion is addressed.

Street tree inventories can provide information invaluable to program planning and subsequent management. In such inventories, it is generally desirable to obtain the following information:

1. Total number of trees.
2. Species composition.
3. Tree location.
4. Size composition.

*For further information on timber cruising, please see *Log Scaling and Timber Cruising*, by J.R. Dilworth, Oregon State University Book Store, Corvallis, Oregon, 1968.

5. Age composition.
6. Condition class.
7. Management needs (pruning, wound treatment, insect and disease control, removal).
8. Planting needs.
9. Total tree values.

It should be remembered, however, that the information from an inventory ought to be the information needed. Thus, it is very important to know what is needed before beginning. If, for example, the need is for tree numbers and estimated total value in order to support a maintenance budget request for a newly annexed area, a "windshield count" of trees by species on alternate streets might suffice. If, however, the need is to develop or improve a total planting, maintenance, and removal program, an inventory should be designed to provide information similar to the above. The key is to know what is wanted from the inventory before beginning. We say this with conviction because we have observed, in our judgment, far too many elaborately designed and expensively made inventories than the situations warranted. Inventories are expensive if they provide less, or more, than what is needed for improved management. We believe, in general, that inventories ought to be conducted by qualified professionals and that on-site judgments should be made of planting, maintenance, and removal needs, rather than recording data such as branch condition, leaf quality, space, and other indicators. In short, why not call it then and there, instead of collecting a great deal of expensive data in order for the same decision to be made, perhaps less accurately, later. In their excellent summary of urban tree inventory systems, Sacksteder and Gerhold (1979) offer the following recommendations:

> Prior to instituting an inventory system, a municipality should carefully consider exactly what information is needed. An excellent guideline for doing this is presented by Ziesemer (1978). It may also be advantageous to consult an expert, whose experience can be invaluable in avoiding pitfalls. When this task is completed, the city may choose to: (1) not start an inventory because it is not feasible now; (2) conduct an inventory without examining systems that have already been developed; (3) use one or more existing systems as models; or (4) adopt an existing system with or without modification. The first choice may be a realistic one, depending on local circumstances. The second choice is obviously not recommended, for there have been too many failures with this approach. Most systems have been developed using a mixture of choices (3) and (4). Lately there has been a definite trend toward using professional consultants, especially in the private sector.

A difficult problem in tree inventories is locating trees for future reference. One solution has been to number trees on a street individually; however, such a system is invalidated when trees are removed or planted. In another system of inventory house or lot numbers have been given to tree locations. This is made difficult by trees on lot boundary lines. Probably the surest system involves distances from street intersections. This fixes the location so that any question as to what tree to prune or remove (etc.) can be resolved by measuring the distance noted. An inventory system we have developed is given in Appendix 3. This system can be used manually or by computer.

A computer inventory system lends itself well to city tree management by providing easy access to data for developing planting plans, planning for dead tree removal, implementing systematic maintenance schedules, and controlling insect and disease situations. A useful aspect of computerized inventory data is that some system of cross tabulation can provide combinations of variables and can enable specific areas of management needs to be mapped. Information such as condition class by species, management needs by species and location, planting needs by size and location, removal needs by location and difficulty class is readily available. In all cases, it can be retrieved very rapidly and with minimal effort.

Based on the information obtained from the inventory, municipal foresters can revise and update street tree management programs. For example, planting requirements can be determined by location, concentration, species, and number. This can be determined citywide, by area, by street, or by block. Proper selections of species can then be made, purchase requirements determined, and planting schedules formulated.

Tree removals follow the same format. Location and concentrations are known as well as removal difficulty class. Thus projections of equipment and man-hours, along with systematic schedules, can be formulated.

Tree maintenance can also be addressed fully. From the inventory, maintenance needs (e.g., pruning, wound repair, insect and disease control) can be determined citywide. Systematic schedules can then be developed and priority areas designated for effectively assigning work crews and equipment.

In larger cities, subsample techniques may be considered to reduce inventory expense. Usually tract development areas are quite homogenous with respect to tree species. Therefore, instead of 100 percent inventories, 10 or 20 percent subsamples of tree populations might be all that are needed to provide adequate information for various projects. Although the specifics for individual trees are lost, all aspects of an overview are retained.

Inventory data can also be used to estimate shade tree values. By using shade tree valuation formulas, values can be produced for the total tree population, for each species, or for individual trees. Such values can be used to stress the importance of a city's tree resources and to support budget requests.

Formulating an annual budget requires factual information. How many trees need pruning, fertilizing, insect and disease control, or removal? What are the total planting requirements? The inventory data can answer those questions. Based on the data, purchases of supplies and materials can be accurately forecast, manpower requirements determined, and assessments made regarding contract and in-service work.

A tree inventory system provides only data. Effective use of the data is the key to budget maximization. Savings in annual operating costs alone may offset inventory costs. Those responsible for municipal forestry programs benefit from efficient operation, the urban forest benefits from proper care, and the residents have a better environment.

A sample city street tree inventory system adaptable for either manual or computer data retrieval is given in Appendix 3.

Planting

Planting is perhaps the most publicly acceptable management need of the urban forest. Most people recognize this need because it is basic, tangible and reasonable. Tree planting is further acceptable since it fosters a satisfying sense of achievement. There is little doubt that we have been influenced by the poets' expression of the nobility of planting trees.

Now planting must have high priority; however, at times utility line clearance, dead tree removal, or other necessary and more expensive practices must take precedence. Thus, there is no contradiction with the earlier statement that budgets for utility line clearance often exceed the amount of money spent for planting, insect and disease control, and all other practices combined. There are three primary management considerations regarding planting in the urban forest: composition, location, and design.

Species composition

Experience with Dutch elm disease has recently brought into sharp focus the need for influencing the composition of the urban forest. The aftermath of the disease in communities where American elms once made up as much as 90 percent of the total tree population is tragic. In communities with lesser percentages of elms, the impact has obviously been less severe. This experience has led many municipal forestry departments to adopt a policy of species diversification so that no single species makes up more than 10 or 15 percent of the total population. The wisdom of such a policy is difficult to question provided that species concentrations and age classes are also considered.

The most direct way of influencing species composition is for all streetside, park, and other public area planting to be done by crews of the municipal forestry departments. The second most direct method is for planting to be contracted with private firms. In both cases, municipal government has control of the planting operation; in fact, this is the situation in most larger American cities. These operations are usually based on long-range master planting plans that consider species, site, and social factors. In smaller cities and towns, streetside planting is often the responsibility or option of adjacent property owners. In such cases, composition is often controlled by less direct means such as planting permits and official species lists. These are prescribed by ordinance but often suffer from inadequate enforcement.

An even less direct method, often used in small towns, is the wholesale purchase of planting stock by a department, board, commission, or other organization. Such purchases are made on the basis of orders by property owners from a selected species list. The trees are sold at or near cost and may be planted by city crews, civic organizations, or property owners themselves. The species list usually includes several choices by size classes (small, medium, and large trees). In this way, composition can be controlled to a degree and spatial factors can also be taken into account. The species offerings are usually changed from year to year for further diversity.

Species composition can be influenced to an extent on private property by ordinances that prohibit the planting of certain species. Such prohibitions usually result from undesirable characteristics such as "cotton" from cottonwoods, messy fruit from certain varieties of crabapples, or the offensive order of ginkgo fruit. Composition can also be influenced by encouraging the planting of certain species; this is done through news media releases by municipal forestry departments, often in cooperation with local nurseries.

Location

There are two aspects to the challenge of locating new trees and related plants in the urban environment: (1) to locate plants for minimum interference with the objects and workings of society; and (2) to locate them for maximum environmental enhancement. This challenge is perhaps best illustrated by the selection and design of streetside plantings. Spatial factors were discussed in Chapter 5. While these factors influence all planting in the urban forest, they are particularly critical to streetside trees and involve potential interference with curbs, sidewalks, driveways, fireplugs, street intersections, traffic signals, signs, streetlights, overhead wires, underground utilities, buildings, and existing trees.

Four general streetside situations can be identified: (1) parkways or treelawns between curbs and sidewalks—these are generally found in older residential areas but often extend into business districts and other areas; (2) treelawns not defined by sidewalks—as in newer residential areas; (3)

sidewalks extending from curbs to buildings—generally in business districts; and (4) streetsides with no well-defined treelawns—mainly in strip developed commercial areas along major streets. The tree planting challenge varies greatly between these situations and warrants further discussion.

Treelawns Between Curbs and Sidewalks:

Rectangular planting was the traditional design of American cities until the early twentieth century. This design was influenced largely by our rectangular system of land survey, which offered a convenient base for further division into blocks and lots. Departures from this design resulted generally only when topographic or other landscape features were too imposing. Sidewalks adjacent to streets filled an obvious need in residential areas in pre-automobile days. It can be argued that sidewalks were placed well away from streets simply to provide safety for pedestrians (or perhaps to protect them from splashing by water and mud) or that they were deliberately so located to provide a lawn for trees. In either case, these areas were heavily planted to trees, which have since competed with utilities, signs, lights, and other objects.

Original trees were often planted in the center of the treelawn with strict regimentation by straight row and spacing (Figure 6.1). Spacing was

Figure 6.1 Parkways, or treelawns, between curbs and sidewalks with trees planted in straight rows are common in older residential areas.

146

most often too close, perhaps reflecting a lack of appreciation for mature size. Close spacing was also a result of narrow lots with property owners often wanting two trees at the front of 40- to 50-ft (12.19- to 15.24-m) lots.

New planting means primarily individual tree replacement, since opportunities for design considerations such as form, texture, color, and scale are limited. Exceptions are areas where extensive losses from Dutch elm disease or other causes have occurred and have prompted large-scale replacements.

Treelawns Not Defined by Sidewalks: Modern subdivisions are products of our automobile-oriented society. They are characterized by wide curving streets, few sidewalks, large lots, and houses set well back from the streets (Figure 6.2). These areas often allow departure from the traditional single row of streetside trees and provide opportunities for informal design. There are generally fewer obstructions to trees, as lawn spaces are larger and utility lines are frequently buried. Tree planting is often controlled by subdivision regulations that require a specific number of shade trees per lot. Such regulations often also prohibit tree planting within certain distances of street curbs. However, in subdivisions that do not have such reg-

Figure 6.2 Modern subdivisions are characterized by wide curving streets, few sidewalks, and houses set far back from the street.

147

ulations, streetside trees are planted by developers and property owners and through planned programs administered by city forestry departments. These programs vary from specific design input by landscape architects to crews being sent out with general spacing instructions.

Design is sometimes governed by maintenance considerations. Maintenance is the primary rationale for lining individual streets with a single species. This practice is based largely on the idea that insects and diseases specific to individual species can be more efficiently controlled. Such plantings are further excused on the premise that repetition of a single species provides unity and harmony. It is also often thought logical to line streets named after trees with that tree. Thus, Maple Street ought to have maple trees along it. The main reason, however, is maintenance even though there is little evidence that maintenance costs are, in fact, reduced. The case against this practice is that site differences are often ignored and that repetition is frequently monotonous. The strongest argument, though, is that if an effective control capability is not maintained, insect or disease epidemics could denude entire streets. It is perhaps unrealistic to assume that such a capability can be maintained throughout the lifespan of the trees.

Business Districts: Business districts, with sidewalks extending from curbs to buildings, are among the most difficult areas for tree planting. Spatial and site problems are most severe and people impact is greatest. If properly located, however, trees can have a very favorable influence on the visual environment of these areas (Figure 6.3). Trees can unify architectural elements, "soften" hard building materials, provide color and character, and create seasonal change and variety. Perhaps no other operation can do more for less cost than the planting of trees.

Streetside trees in business districts must be carefully located. They should not be used in front of major store entrances or display windows but may be used in front of areas between windows or where buildings join. Trees must also not obscure advertising signs or marquees. Automobile parking arrangements must be considered. For angle parking, trees must be located a minimum of 30 in. (76.2 cm) from curbs to prevent damage by car bumpers. They should also be located directly in front of parking spaces so that they do not interfere with pedestrian movement between cars. Obviously above and below ground utilities, parking meters, and other physical features must also be considered.

If careful attention is paid to all site factors, trees can often be planted in sidewalk wells. This method is generally preferred in that greater soil volume is available for root development. In situations where underground obstructions prohibit sidewalk wells, containers and raised beds are often used. Design, location, species selection, and maintenance are extremely

Figure 6.3 Trees can be grown successfully in business districts and can have a very favorable influence on the visual environment.

important. Restricted soil volume is the main problem and frequent watering is called for. Raised planters are also convenient places for litter disposal, and daily cleanup is often necessary. If properly designed and a strong commitment made to maintenance, containers and beds can support trees and other plants beautifully. If not, they will rarely be successful.

Species and variety selection are very important as trees must conform to available space. Trees with upright form are most commonly used as there is limited space for lateral branch spread. Selection and location must be done by competent personnel fully cognizant of all factors. Such work is often done by qualified landscape architects. In some cities such projects are funded by store owners, business organizations, or other sources. Generally, however, business district planting is the responsibility of city forestry departments.

Strip Developments: Commercial strip developments along major streets are extremely difficult planting situations. Such areas are often a profusion of driveways, signs, poles, and wires badly in need of trees to enhance the visual environment but with few logical places to plant them (Figure 6.4).

149

Figure 6.4 Strip developments are badly in need of trees to enhance the visual environment, but there are few logical places to plant them.

Commercial considerations simply take precedence over trees along such streets. As with business district plantings, proper location is extremely important, requiring that all factors be considered.

Design*

The street corridor must be considered a volume or a space and not simply an elongated or lineal ground plane. A successfully designed streetside landscape will be open where pleasant views or safe vision is desired; closed where visual screen is needed; and varied in form, size, texture, and color for interest. Spatial variety is important in preventing driver fatigue, maintaining driver alertness, and emphasizing danger zones. It also helps in developing a satisfactory visual-sensual experience. Spatial variety must, however, be properly designed. Too much variety leads to disorder, a lack of harmony and continuity, while too little variety results in monotony. The ideal street corridor will be scientifically engineered and artistically formed of expanding and contrasting spaces in which adequate variety exists. Mo-

*These sections on design, pp. 129–139, were prepared by James J. Nighswonger, Landscape Architect, Department of Forestry, Kansas State University.

torists and pedestrians should move safely and freely through the corridor, enjoying a streetscape designed to keep them relaxed, happy, and alert.

A street corridor space, whether occurring naturally or intentionally designed, will be enclosed by base, vertical, and overhead planes (Figure 6.5). It is a three-dimensional volume limited only by the relationships of these three elements of space enclosure. In the street corridor the space is usually experienced while driving, riding, or walking. To the viewer it will be a continually changing space primarily affected by changes in the vertical plane.

The base plane within the street corridor space is usually consistent in color and texture within a given area. It may be comprised of concrete, asphalt, brick, gravel, grass, or combinations of the same. Lack of variety in this base plane induces monotony, boredom, and fatigue. The overhead plane likewise is not readily perceived as variable when comprised of sky or cloud cover. More variety and interest exist, however, when this plane is formed by a broken tree canopy that creates light and shadow patterns within the corridor and a more highly defined enclosure.

The most variety and interest occur within the vertical plane. This portion of the three-dimensional enclosure of space is also the area where trees and shrubs have their greatest influence. Therefore, street trees, buildings, walls, solid fences, vertical land forms, and other objects that create vertical enclosures are the most significant parts of the space. In the creation of this vertical enclosure the importance of the proper use of street trees with respect to form, size, texture, and color cannot be overemphasized!

Designing the vertical enclosure

The vertical enclosure created by vegetation or other objects provides visual control. Everything that occurs within the enclosure is a part of the visual function of that space and must be taken into account. A desirable object can be emphasized and, conversely, an unattractive object negated by manipulating the vertical enclosure (Figure 6.6). Strong contrasts in form, size, texture, and color, or combinations of the design elements will create interest and lead the viewer's eye to a desired object (Figure 6.7). By the same token, repetition of any one of the design elements may tend to negate an associated object (Figure 6.8).

An object outside the vertical enclosure of the street space can be introduced into the space visually. A distant mountain or meadow, a lake, a building, or any other pleasing feature can become a part of the visual scene when openings in the vertical enclosure are designed that permit viewing and provide enframement of the outside object.

Generally strong contrasts within the vertical enclosure should be avoided. Repetition and subtle changes in form, size, texture, and color are desirable.

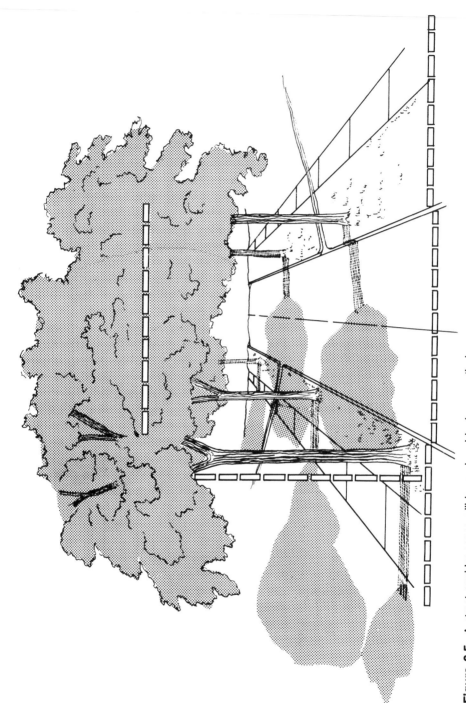

Figure 6.5 A street corridor space will be enclosed by base, vertical, and overhead planes.

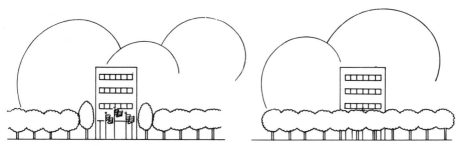

Figure 6.6 Manipulation of the vertical enclosure can emphasize or negate an object.

Figure 6.7 Strong contrasts in texture and color will attract viewer attention to a desirable object.

Figure 6.8 Repetition of the design elements, (form, size, texture and color) will negate an associated object.

Exceptions to this principle occur at major street intersections or at any other area where alertness and viewer attention are desired (Figure 6.9).

Design elements

Street trees when properly used perform limited engineering functions such as sound control and filtering polluted air; their architectural functions such as softening building lines provide a visual screen or privacy; and their climatological functions provide shade and wind control. Any of these purposes alone may be reason enough to plant trees along street sides. In reality, trees provide a combination of these functions in any setting, thus contributing substantially to human comfort and enjoyment.

However, in order to select and design street tree plantings properly, esthetic functions and art principles must receive consideration. The terms such as sequence, repetition, rhythm, unity, emphasis, and scale are basic to good street tree design. Also the four elements—form, size, texture, and color with respect to tree characteristics—must be understood by the designer.

153

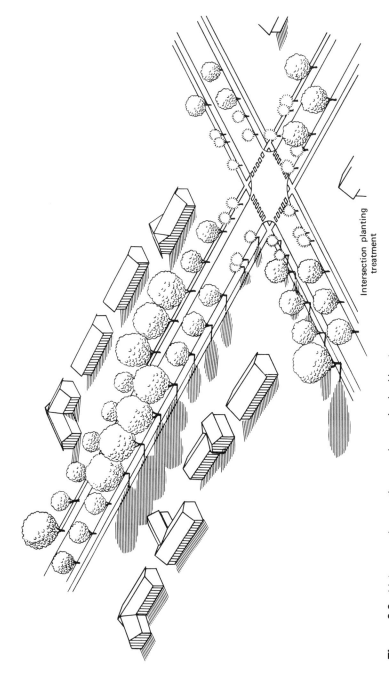

Intersection planting treatment

Figure 6.9 Major street intersections can be emphasized by strong contrasts in the vertical enclosure. This intersection is emphasized by abrupt changes in tree size, color, and texture.

154

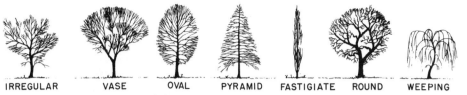

Figure 6.10 Major characteristic tree forms that occur in natural and cultivated species.

Form: All trees have a characteristic form or shape under normal growing conditions. The form of some varieties changes as a tree develops from its juvenile size to maturity; however, the designer should be most concerned with mature form since that is the characteristic form. A tree's form is created by its outline, branch and twig structure, and habit of growth. Form is such a prominent characteristic of many species that a tree can be identified from it often from several blocks away (Figure 6.10). Of the four tree characteristics, form is the strongest design element and must be considered in any well-designed group planting or streetside landscape. The most suitable forms for streetside situations include round, oval, upright oval, fastigate, or irregular as compared to pyramidal and weeping forms. Pyramidal or weeping forms are seldom suitable in streetside applications. Such trees occupy space often needed for vehicular and pedestrian movement (Figure 6.11). They also create visual obstruction and will dominate in the streetside landscape when they are foreign to the characteristic landscape.* Pyramidal forms accent, especially when used in association with rounded, oval, irregular, or weeping shapes. For these reasons trees assuming a characteristic pyramidal form are best used in park or large open-space development or, in some instances, in wide boulevard plantings where their low branching habit and density can reduce headlight glare and provide a visual screen without occupying the physical space needed for vehicles, pedestrians, or visual access.

Size: All trees, given normal growing conditions, can be expected to attain a certain ultimate size. As a design element, size is probably next in importance to form—and size is definitely the most misused element in streetside landscapes.

The layperson will frequently select a tree on the basis of how well he likes a certain species and give little regard to how large it will grow. Thus we commonly observe trees growing into overhead utility lines, breaking

---

*For our purposes the term "characteristic landscape" refers to the predominate environmental character of a specific area as influenced by a particular tree character and composition that dominates.

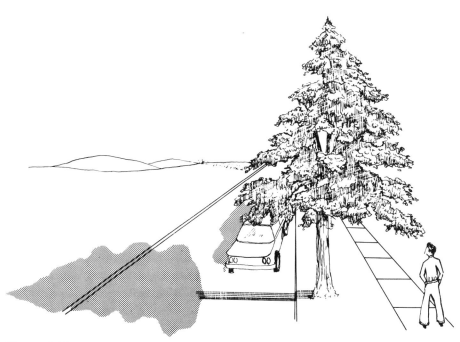

Figure 6.11 Trees with pyramidal or weeping forms may occupy the physical and visual space needed for the safe movement of vehicles and people.

sidewalks and curbs, obscuring views, creating traffic hazards, and growing out of scale with respect to their surroundings.

For design purposes, trees can be grouped into three size categories: small, medium, and large, as discussed in Chapter 5. These are illustrated in Figure 6.12. In the past, communities throughout much of the country had streets that were lined and canopied by large elms, oaks, maples, and other species. This concept often resulted in beautiful, well-defined spaces that many people fondly remember (Figure 6.13). The modern-day street scene has changed, however. Driving surfaces on streets and boulevards are much wider; utilities both above and below ground are more numerous; curb cuts, drives, and sidewalks occur more frequently, all resulting in reduced space for trees (Figure 6.14). Consequently large growing trees seldom have a place along the street today. Even medium trees grow too large for some situations. Small and medium growing trees should therefore comprise the basic landscape in most modern street corridors.

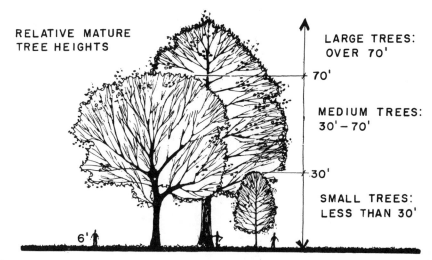

RELATIVE MATURE
TREE HEIGHTS

LARGE TREES:
OVER 70'

70'

MEDIUM TREES:
30' – 70'

30'

SMALL TREES:
LESS THAN 30'

6'

Figure 6.12 Size categories of mature trees for design purposes.

Figure 6.13 Streets canopied and lined by large trees are attractive, well-defined visual spaces.

157

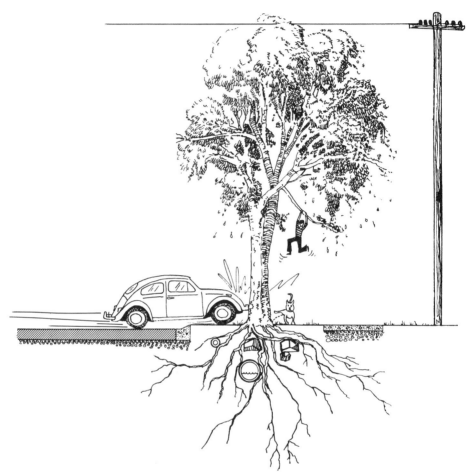

Figure 6.14 Competition for space along the modern-day street often leaves little room for trees, especially larger sizes.

Texture: The term texture, when describing tree characteristics, refers to visual and not tactile texture. The concern is thus for the visual texture of a tree or a group of trees and the relationship to other plants in the same visual space. Texture may be described as coarse, medium, or fine (Figure 6.15). However, this description is relative. A plant's texture can best be described as coarse when compared to another plant that possesses a medium or fine texture and vice-versa.

A tree's texture is created by foliage and twig size primarily, with bark characteristics also playing a part. Since foliage size is the most important

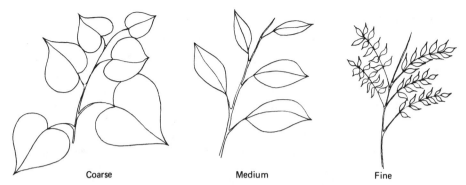

Figure 6.15 Relative foliage textures as perceived visually.

factor, texture relationships with deciduous trees are most significant during the growing season.

When considering texture in a plant composition, uniformity and continuity are important. Subtle change in texture is appropriate and will add interest to the street space. On the other hand, abrupt textural changes will accent or create dominance; thus abrupt change should be designed only when dominance is desired. For example, in a grove of honeylocusts (fine texture), a catalpa (coarse texture) will stand out or dominate. Likewise a sycamore, London planetree, or basswood will dominate when used with fine-textured trees as part of a design.

Color: Color is fourth design element. Again, uniformity with subtle change is important when considering color combinations. Various shades of green predominate the summer scene in any tree grouping. The shade or tint difference will vary with species and varietal changes, health or physical condition of the plant, soil or nutrient change, age of the foliage, and other factors. All greens that occur in nature generally blend well with one exception; yellow-green foliage may dominate if its intensity is strong. Generally the designer should avoid blues, purples, reds, and other colors when they are not common in the characteristic landscape. Plants assuming such contrasting colors will attract attention. As before, if the purpose is to accent or attract attention, then color contrast is another device available to accomplish this objective.

Fall colors—reds, browns, yellows, and greens—often blend well with one another because they are related or analogous. High intensity can still result in dominance even in a related color, however. A knowledge of complementary and analogous colors, intensity, shades, and tint is important when designing with color, whether working with paint or plants.

159

Putting it all together

The use of form, size, texture, and color in the development of esthetically pleasing and functional spaces is the key in street tree design. Many of the factors involved have been discussed in this chapter; however, some additional comments should be made.

Seldom will the designer utilize all four of the design elements at one time in concept development, but all four must be considered. To achieve design continuity, repetition of elements is basic. For example, careful selection and spacing of trees with respect to size, form, and color, or size, form, and texture are common approaches. The common denominators should usually be size and form (Figure 6.16).

Also, it is best to utilize plants in odd-numbered groupings, for instance, in groups of 3, 5, 7, or 9, when informality or "naturalness" is desired in a design.

Exceptions to the above guidelines occur when emphasis or viewer attention is desired. In such cases, the more variety or change from the characteristic landscape the stronger the emphasis. For example, in a composition where medium-sized trees, round in form, green in color, and fine in texture predominate, plants that are large, pyramidal, bright red, and coarse will seem to shout out for attention. In addition, the use of even numbers of dominating plants at both sides of a desirable view or dangerous area where viewer alertness is important will create even stronger emphasis.

Maintenance

From a management perspective, maintenance of the urban forest may be defined as the implementation of practices necessary for reasonable health, vigor, and compatibility with the urban environment. Maintenance involves all practices between planting and removal. These practices can be grouped into three categories: (1) growth control; (2) damage control; and (3) insect and disease control. There is some overlapping of these categories. For example, certain growth control practices can prevent damage and influence disease control. This overlapping is considered in the following discussion. We will discuss maintenance generally from the standpoint of private property owners and more specifically of municipal governments.

Growth control

There are two types of vegetative growth control practices in the urban forest: those that retard or redirect growth, and those that enhance growth. The first category primarily involves pruning, although some work with

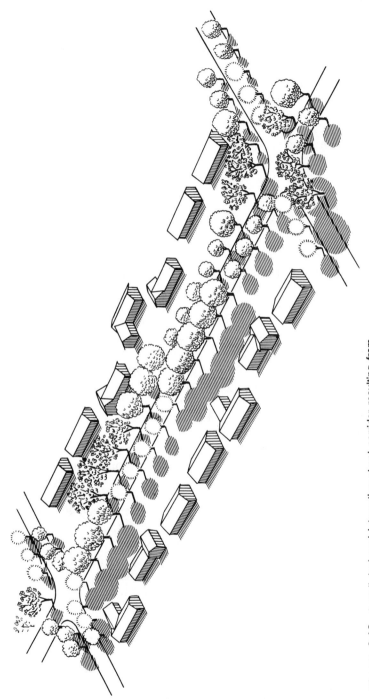

Figure 6.16 A well-designed interesting street corridor resulting from repetition without monotony, and subtle change in size, form, and texture.

chemical retardants is also included. In the second category are practices such as irrigation, fertilization, and control of competing vegetation.

Pruning: Pruning is one of the most important management practices in the urban forest. It may be done for the following reasons:

• Reduction of hazards to life and property.
• Clearance of utility lines or other objects.
• Development of structural strength, shape, and form.
• Appearance.
• Production of fruit.
• Exposure or enhancement of vistas.

Proper pruning is a specialized practice requiring knowledge of plant growth and response. This knowledge is frequently lacking, however, and one can find abundant examples of improper pruning in virtually every city. Pruning is often thought to entail only the removal of lower tree branches, or worse, the near complete removal of tree crowns. Topping, or modified pollarding, is most often practiced in the central and western United States, particularly where Siberian elms abound. Topping is the indiscriminate removal of a tree's crown, leaving large branch stubs vulnerable to decay and resulting in a profusion of adventitious branches (Figure 6.17). The reasons for such improper pruning are based largely on the fear that during storms tall trees will fall on houses or other property. Topping has been done often enough also so that it appears to be the correct thing to do. However, it is a tragedy of many cities and towns that large sums are expended on topping in this misguided belief. This belief is often encouraged by contractors who either know no better practices or find topping the most profitable. This indictment of topping reflects our strong feelings and does not obscure the recognition that it may be necessary as a last resort in cases where trees have suffered severe storm damage. In such cases, detrimental effects can be reduced by making slanting cuts with branches or buds at the peak. As a normal practice, however, topping is simply not justified.

Pruning of street and park trees may be contracted to commercial arborists or done by municipal crews. In many cities where forestry program budgets are limited, priorities for pruning are necessary. These priorities are usually established according to area. For example, pruning of trees in a centrally located park would probably be more important than pruning of streetside trees in an industrial area. Priorities must also be based on considerations of safety, tree value, species, and size. Pruning for safety is obviously needed to eliminate hazardous branches, to clear sight lines for traffic, and to reduce cover for criminal activities in parks and other areas.

162

Figure 6.17 Topping is the indiscriminate removal of a tree's crown. Tragically, large sums of money are expended on this destructive practice in the misguided belief that it is the proper thing to do.

There is little doubt that much pruning in the urban forest is made necessary by the improper selection and location of trees. There are, however, few trees that do not require pruning at some times during their lifespan. Pruning of young trees to develop structural strength and good form should be given a high priority; this can prevent more expensive pruning much later.

In spite of the wishes of many people, the placement of all power lines underground is unlikely and line clearance will continue to be an important and expensive pruning job in the urban forest. Line clearance is usually the responsibility of power companies or municipal utility departments. Clearance specifications are prescribed for various types of lines, and tree limbs and other obstructions must be removed. Utility line easements allow entry on private lands for line clearance and other maintenance. City ordinances and easement regulations often prohibit tree planting beneath wires or prescribe certain low-growing species.

Proper utility line clearance requires a knowledge of tree species, growth characteristics, and response to pruning. Directional pruning is often practiced, particularly for species with ascending branches. Paths for lines can be cleared through trees or branches can be trained to grow away from lines

163

Figure 6.18 Paths for utility lines can often be cleared through trees, and branches can be trained to grow away from lines.

(Figure 6.18). Such methods are not practical, however, for trees with strong terminal branching habits, and drop crotching and topping may be the only recourse (Figure 6.19). The quality of utility line clearance pruning has increased in recent years largely as a result of environmental awareness, corporate image concern, and efforts of the Utility Arborist Assciation, an affiliated association of the International Society of Arboriculture.

In an effort to reduce costs of manual pruning, considerable work has been done with chemical growth inhibitors. Such chemicals either directly

Figure 6.19 Trees with strong terminal branching habits, such as pin oaks, offer little recourse other than topping or drop crotching under utility lines.

prevent cell division in terminal buds or influence hormones that affect growth. Some growth-retarding chemicals are currently on the market, and these products are useful under certain conditions. However, much research remains to determine the practical application of "chemical pruners."

Growth Enhancement:
Increasing growth by practices that maintain plant health and vigor is a necessary management need of the urban forest.

165

These practices may directly or indirectly influence plant growth and may involve irrigation, fertilization, soil aeration and amending, thinning and release, and grass and weed control. The intensity of application of these practices depends largely on the functional value of individual trees or groups of plants. There is often a conflict between growth enhancement and the need to restrict growth. This conflict can be moderated if health and vigor can be enhanced, and increased growth can be redirected.

Damage control

Control of tree damage involves prevention and repair. Much damage can be prevented by judicious pruning and, as suggested in the discussion of topping, much damage can be done by improper pruning. Damage prevention involves reducing the likelihood of physical damage and preventing of wood decay. In addition to pruning, these measures include construction and grade change precautions, cabling and bracing, wrapping, treatment of pruning cuts, mulching to discourage damage by lawnmowers, construction of physical barriers, posting of signs, and policing. Damage repair includes the treatment of cavities, breakage, and other wounds. The objective is to prevent decay and put wounds in the best condition for closure.

Damage control must be integrated with all management practices as it involves correct location of trees, proper pruning, maintenance of health and vigor, and insect and disease control, as will be discussed. It also must involve information, education, and enforcement by responsible officials.

Insect and disease management

Many insect and disease situations can be tolerated as a natural part of the urban forest environment. A point at which a particular outbreak can no longer be tolerated is difficult to determine, both from the standpoint of reduction of health, vigor, and function of the trees, and its annoyance to people. The difficulty is compounded since often control measures must be preventive and initiated well in advance of the actual damage. Effective control involves several elements: (1) knowledge of insect and disease life cycles; (2) monitoring; (3) decision making; (4) physical capability for control application; and (5) legal authority for action. These ingredients must operate also within the climate of public attitudes concerning the use of pesticides. Fortunately, for control purposes, most insects and diseases are specific to individual tree species or varieties. Occasionally, however, insects are less discriminating, particularly when they are at high population levels. Recent gypsy moth outbreaks in Eastern states attest to this.

A basic management approach considers the tree insects and diseases that can be controlled on an individual tree basis and those that require control measures applied to the entire tree species population. Thus, the

166

critical question is whether a particular tree can be protected within tolerable limits by control measures applied to it alone. If individual trees can be protected, priority systems can be set up for control application. Obviously, such decisions must also consider the impact of insect or disease attacks on the health and vigor of individual trees.

Legal authority for insect and disease control programs has been clearly established in most states. State statutes authorize municipalities or other local units of government to levy taxes to pay for control of epidemic pests and diseases and to allow enactment of local ordinances to carry out such programs. The exercise of such authority is perhaps best illustrated by efforts to control Dutch elm disease.

Dutch elm disease is a fungus affecting the water-conducting vessels of elm trees. It is spread primarily by elm bark beetles but is also transmitted by root grafts of adjoining trees and to a minor degree by unsanitary pruning tools. The main vector, however, is the European bark beetle, and most control measures have been directed toward reduction of beetle populations. Recent developments of soluble fungicides and systems for their applications offer promise for a more specific control of the fungus. Elm bark beetles breed in dying elm trees and infect other elms mainly by feeding in twig crotches. Control measures are spraying, sanitation, and prevention of transmission through root grafts. Spraying involves the application of insecticides to destroy the feeding beetles. Sanitation is the removal and destruction of breeding wood. Prevention of transmission through root grafts is primarily by chemical barriers. To be effective, Dutch elm disease programs require: (1) adequate public financing; (2) legal authority and systems for the designation and condemnation of breeding trees on all public and private property; (3) physical capability for timely spraying and tree removal; and (4) a continuing commitment. Numerous cities have demonstrated that as long as the program was diligently followed, elm losses could be held to a very small percentage of the total population. A continuing commitment is extremely difficult, however, in a climate of tight budgets, environmental concerns, and public attitudes. Cannon and Worley (1976) found that some communities temporarily suspend control programs during a period of temporary financial stress only to find that they could not regain control of the disease situation later. They also observed that the communities that experienced the fewest elm losses had well-founded control programs, conscientiously applied, and sustained over the years.

Concern over widespread use of chemical sprays plus the high costs of materials and applications have prompted increased interest in alternative methods of insect and disease control. Integrated control programs are being developed. A model program, developed by William and Helga Olkowski of the Department of Biological Control, University of California, Berkeley,

MANAGEMENT OF THE URBAN FOREST

has received widespread attention. In effect in the city of Berkeley, this program considers the management of insects and diseases as the management of ecosystems. An intensive study of problem situations is required. Often insect problems can be solved by biological controls involving the introduction of natural enemies or the creation of conditions favorable for natural enemies. Certain leaf-eating caterpillars are safely and selectively controlled by spray applications of *Bacillus thuringiensis,* a bacterial enemy of lepidoptera larvae. Additional treatments are water sprays and selective pruning to physically remove insects and sticky adhesives to trap insects. Chemical sprays are also used but in greatly reduced amounts, a result of careful monitoring and optimum timing of applications. A strong effort has also been made to inform citizens of the program. Operational in 1972, the program has resulted in lower pest management costs, fewer citizen complaints, the elimination of secondary pest outbreaks, and a reduction of environmental contamination (Olkowski and Olkowski, 1975).

As indicated, successful pest management requires understanding the biology of particular pests as they interact with trees or other plants to be protected. Underlying most pest control failures is the use of a control method without understanding it. Mock (1982) lists several problems associated with sole reliance on chemical control attempts:

There are no satisfactory pesticides available to control some pest species.

Proximity to dwellings, livestock, bee hives, or water resources may make safe pesticide application impossible.

Adverse weather conditions may prevent application or reduce the efficacy of pesticides that are applied.

Inadequate coverage, poor placement, or improper timing.

"Insurance" applications made without real knowledge of specific pest populations present.

Attempts to subdue outbreaks that have progressed to an uncontrollable stage.

Pests develop resistance to pesticides.

Destruction of natural enemies of pests.

Outbreaks of secondary pests.

Health hazards to users and people who come in contact with treated areas or products.

Thus, Integrated Pest Management is based on a thorough understanding of pests and their influence so that biological, cultural, mechanical, and chemical techniques can be used to support one another.

168

The extreme complexity of urban forest pest management is perhaps best illustrated by efforts to control gypsy moth outbreaks in Eastern states. The program of the Monmouth County, New Jersey Shade Tree Commission is a model example. Cooperative with the New Jersey Department of Agriculture and the USDA Forest Service, the program provides the following services: (1) surveys to determine the size and location of infested areas; (2) release of biological parasites in infested areas; (3) evaluation of the effectiveness of parasites; and (4) financial assistance to communities or municipalities for spraying, contingent upon legislative approval. Communities or municipalities must then: (1) request the above services; (2) declare the Gypsy moth a "public nuisance"; and (3) notify residents, especially beekeepers, chicken farmers, and others, of proposed spray programs. Included are innumerable details of education, communication, and coordination. The Monmouth County Shade Tree Commission (Shaw, 1981) reported receiving over 6,000 telephone calls during the ten weeks of larval infestation. They also held meetings with local gypsy moth coordinators to insure that proper notification was given to people in designated spray areas, conducted field trips on biological aspects of moth control, made honeybee pollen traps available to beekeepers, surveyed all areas where moths were reported, gave daily radio reports, and conducted educational programs through schools and various organizations.

Fire protection and control

Fire protection and control in urban areas is generally not a function of municipal forestry departments. This responsibility is usually vested in separate fire departments and has as its objective the protection of life and property. Thus, the protection of the urban forest for itself is secondary and is of major importance as it influences life and property. This is not to say that urban forest values are not considered in fire protection and control; they are simply secondary to life and property.

Under certain conditions, all forests are subject to fire, and with potential in many cases for disaster. This is particularly the case in natural forested conditions of highly volatile fuels where urban development has occurred. The most common example is in the Los Angeles area where the urban forest includes literally thousands of acres of chaparral and associated vegetation. This vegetation, plus topographical, weather and population-related factors, has resulted in near-annual conflagrations of disastrous proportions. Fire protection under such conditions is highly complex and involves such things as land use planning, zoning, fire code development and enforcement, public education, vegetation control, and maintenance of a highly sophisticated fire department. The Los Angeles County Fire Department has a Forestry Division with responsibility for: (1) protection of county watershed areas; (2) improvement of vegetative cover; (3) provision of conservation education; and (4) prevention and control of wildland fire (Radtke, 1978).

Most of the elements of urban forest fire protection are present in this area and include such items as enforcement of regulations concerning roof shingles, clearance of chaparral around buildings and replacement with fire retardant vegetation, production of nursery stock for pre-fire and post-fire planting, research on fire retardant plants and their establishment, and public education. The program also involves cooperative agreements and projects with other government agencies and private organizations for soil erosion control, wildlife habitat improvement, vegetation maintenance, chaparral ecology, and fire suppression.

The Los Angeles area is exceptional because of the nature of urban forest fuels and other factors, and it illustrates the complexities at their extremes. As suggested, however, most managers of urban forests are not directly responsible for fire protection and control. Therefore, there must be cooperative agreements with local fire protection agencies. Involvement by the urban forest manager may be extensive or intensive, depending on hazards and values involved. In many cases, urban foresters serve as technical advisors to fire departments on matters such as fuels, fire lanes, vegetation control and establishment, and training.

Contracts

Many municipalities contract with private firms for certain aspects of urban forestry work. In a survey by Giedratis and Kielbaso (1982), 700 cities reported that they use some type of contracted services: 57 percent contracted for tree removal; 54 percent for stump removal; 37 percent for pruning; 30 percent for planting; 22 percent for spraying; 8 percent for storm cleanup; and 7 percent for other services. There appeared to be little relationship to city size and contracting, except for planting, with larger cities contracting much more often than smaller ones. As indicated, tree removal is most often contracted, perhaps due to the specialized equipment needed, and perhaps also because contract specifications are relatively uncomplicated. This same study also showed per unit contracted costs for all services to be generally higher than in-house costs. The authors caution, however, that cities may not have considered all expenses in their reported costing and that contractor's profits are included.

As with all contracts, urban forestry work contracts must have the elements of mutual assent, valid consideration, legal capacity, and legal subject matter, and should be recorded in writing. Most contracts are preceded by a notice to bidders, or an "invitation to bid," containing instructions and specifications. Hoefer (1983) suggests that such notices contain the following:

Instructions

Who is requesting bids.
Where to send bids.
Deadlines (for bid submission and work completion).
Earnest money, or bid deposit.
Bonding requirements.
Arborist or other urban forestry-related license requirements.
Insurance requirements (public liability and property damage).
Instructions for bidding.
Penalties for failure to perform.
Liability disclaimer for the contractee.
Right of bid rejection statement.
Date, time, and location of pre-bid meeting.
Date of bid opening and procedure for notifying successful bidder.
Equal Opportunity and Affirmative Action requirements.
Other provisions as required by contractee.

Specifications and Procedures

Definition of work (what is to be done and how it is to be done).
How work will be coordinated.
Inspection process.
Safety procedures.
Operational procedure, including debris cleanup and disposal, noise limits, hours of operation, and public relations.
Billing and payment policy and procedure.

Although written several years ago, the following 12 points concerning bid specifications and contracts are no less valid today (Plinkerton, 1968):

1. Bidders should be required to carefully examine the site of the proposed work and judge for themselves as to the nature of work to be done and the conditions under which it has to be accomplished. A proposal from each bidder should be requested so that it is assured that he has accomplished this task and understands the character, quantity and quality of the work to be done.

2. Bids should be submitted in a sealed envelope, opened in public and read at a prearranged time. All bids should show the proposed prices clearly, legibly and must be properly signed by the bidder.

3. Bidders must be competent and capable of satisfactorily performing the work covered by the proposal. They should be able to furnish references

171

as to similar work experience, the plan or procedure proposed to accomplish the work and the organization and equipment available to accomplish the task.

4. The award of the contract should be to the lowest responsible and qualified bidder whose proposal complies with the requirements. However, the right to reject any and all bids should be reserved to best serve the interests of the city.

5. The contractor should understand that he is required to perform and complete the work in a thorough and workmanlike manner and to furnish and provide all the necessary labor and equipment to accomplish the task.

6. In the case of a cash contract, the city or agency should reserve the right to make an increase or decrease in the quantity of work to be performed and to correspondingly increase or decrease the amount to be paid to the contractor.

7. The contractor should be required to clean the site of rubbish or debris caused by his activity at the end of each working day and leave the site in a neat, orderly, and presentable condition.

8. All work to be performed should be subject to the direct supervision of the city forester or duly appointed official and should meet with his approval concerning the provisions and requirements of the bid.

9. The contractor should be required to carry public liability insurance as required by the city and to furnish a 10-day cancellation notice clause certificate to the city or agency in charge. In addition he should be required to secure, maintain in full force and effect, and bear the cost of complete workmen's compensation insurance for the duration of the contract. It should be understood that the city or agency will not be responsible for any claims or suits relative thereto.

10. The contractor should make adequate provisions to accommodate normal traffic over the public streets during the duration of the work so as to minimize public inconvenience. This includes ingress and egress of occupants of private property and the furnishing of the necessary barriers, guards, and lights to protect the public.

11. The contractor should be aware that he is responsible for the preservation of all public and private property in the work area and should exercise due caution in his operation to include protection of plant materials, street signs and lights. He should further understand that he will be required to bear the expense of correcting such damage and repair it in an acceptable manner if it occurs.

12. At the completion of his work the contractor should notify the responsible public official that he is ready for final inspection. When this inspection is completed and the work passed, payment can then be made.

In addition to a written list of bid specifications, bidders should be given an idea of what a finished product is expected to look like. The contract should be awarded after weighing carefully the price, reputation, experience, and physical ability of the contractor to accomplish the work in the time specified. Prior to starting the job, it would be wise to meet with the contractor and review all of the conditions including requirements for traffic control and public safety. A sample call for bids can be found in Appendix 4, which illustrates how a bid is worded to cover the 12 points listed previously.

The complexity of urban forestry work contracting is further illustrated by the following considerations important to specifications for landscape planting contracts (Nighswonger, 1981):

Scope of Work

Plant Materials
   Standards (AAN, 1973)
   Plant list
   Substitutions
   Quality
   Measurements
   Preparation of plants
   Delivery
   Inspection

Installation Materials
   Guying
   Soil amendments

Planting
   Time of planting
   Layout
   Setting plants
   Backfilling
   Guying
   Mulching
   Watering
   Pruning and repair

Maintenance

Inspection

Guarantees and Replacements

Removal

Removal (which includes harvesting, utilization, and disposal) of materials from the urban forest is an important management need. Materials that must be removed are dead trees, hazardous trees (physically hazardous and those that harbor potentially epidemic insects and diseases), overcrowded trees, pruning debris, storm debris, stumps, leaves, and obnoxious fruits. Removal also includes the planned harvesting of wood products such as pulpwood and sawlogs from some areas of the urban forest. Depending on the situation, such harvesting can be totally consistent with other uses of the forest. A classic example is in Goldengate Park in San Francisco where the older trees are being systematically harvested and new ones established in order to create an uneven-aged stand—all according to a prescribed management plan and accomplished in such a way that park users are scarcely aware of the loggers' presence. Another example is Philadelphia's John J. Tyler Arboretum where a 90-acre tract is under intensive multiple use management including selective harvesting of sawlogs, veneer logs, and fuelwood.

As with other management practices, the removal of materials from the urban forest must be authorized, planned, and funded. Plans must be flexible enough to deal with emergency situations such as storms or other disasters. Authority for material removed is generally based on police powers that provide for the safety and convenience of residents. Most removal work can be planned. Plans are based on inventories, inspections, or past records.

In many cities, periodic inspections are made of all public trees; and dead, dying, and hazardous trees are designated and recorded. In cities having Dutch elm disease or other disease control problems, dead and hazardous trees on private property are also designated. Owners are notified and are given a reasonable time for removal. In most situations, if such trees are not removed within the designated time, they are removed by the city and the cost charged against property taxes.

The bulk and weight of wood materials to be removed often necessitates on-site processing. Twigs and branches can be conveniently reduced by chipping. Larger portions of tree trunks can also be chipped but require bigger and more expensive machines. Wood chips may be used in the manufacture of particle board, roofing materials, and paper products. They also make excellent mulching materials and are often used as animal bedding and poultry litter. The search for such uses has intensified because laws passed in recent years have prohibited the burning of wood wastes for air pollution reasons.

The problem of disposal of forest debris is illustrated in Chicago, Illinois where approximately 295,000 elms with a minimum trunk diameter of 20 in.

died of Dutch elm disease between 1968 and 1978. Removal costs amounted to about $24 million during this period. In an effort to utilize elm wood and partially offset the cost of removal, the Bureau of Forestry purchased a chip-harvesting machine and arranged to supply chips to a roofing shingle manufacturer, with the expectation of chipping 50 percent of the debris. Also, in order to establish markets for the remaining debris, log contracts were initiated with six sawmills and two pulpwood contractors. The results were not encouraging. Costs so far exceeded returns as to make the operation virtually impractical. Inhibiting factors were (Geiger, 1978):

High percentage of debris with imbedded metal, concrete, crotches, limb stubs, rot, and other problems.

Debris is unsorted when removed from the street. Brush is mixed with logs and usable logs are mixed with usable logs.

Best logs are capable of producing only low value wood products.

Continued inflation of labor and equipment has not been offset by similar increases in revenue for wood products.

The random removals result in extreme variability of debris that must be utilized.

This experience, plus an intensive study, led Chicago foresters to the conclusion that in order to achieve an economically sound and environmentally attractive tree debris disposal program, the city should: (1) implement a short-range wood fuel option with private operators and (2) study the long-range potential of incorporating tree debris in a total solid-waste energy recovery program.

This experience also led Chicago urban foresters to an interesting concept—certainly not original as it has been practiced in less wood-rich countries for centuries—and perhaps a future reality in the United States. Why not develop for the urban forest (streetsides, parks, and other areas) a sustained yield system in which not only dead trees but mature trees are harvested as well? "Shade" trees would be selected, planted, and cultured for ultimate harvest for wood products, after producing interim amenity benefits. Such a system might well produce revenues from harvested products that would pay for forestry program costs. This concept is based largely on the logic that maintenance costs increase sharply as trees become overmature and eventually have to be removed anyway. Why not remove them at the optimum time when dollar value can be maximized?

The obvious problem with this approach is that large old trees have a greater value (utility) to residents by their presence than the value of their products. Many people would view the sight of large live trees being cut down as offensive, and some would find it intolerable. Another problem is that some of the same inhibiting factors as listed above for the recent Chicago experience would probably still apply. Still, it is a concept that should not be discarded out of hand. The logic is sound, and the sheer numbers of our population may one day make it practical.

While the Chicago experience pointed out the problems inherent in attempting to utilize urban trees by conventional methods and the limited opportunity of such methods for offsetting removal costs, their program was by no means a failure. Their dead elms were removed and disposed of, and their program did point the way toward better practices, particularly with regard to fuelwood use. It is in this area that most cities are finding the greatest opportunities for urban forest product utilization.

The utilization of wood materials, particularly debris, from the urban forest must be considered in context with other wood waste, since it is the total of all usable wood waste that largely influences its practicality of utilization. An estimated 60,977 million air-dry tons of urban wood waste is generated annually in the United States (Carr, 1978). Of this, 73 percent is wastepaper, 22 percent is waste timber products (boxes, crates, discarded furniture, wood milling residue, and the like), and five percent is directly from trees. While obviously a relatively small percentage of the total, urban tree waste still amounts to 2.8 million air-dry tons per year, making its disposal and/or utilization a quite important matter. In their lead paper to a conference on "Alternatives to Urban Waste Wood Disposal," Cordell and Clements (1979) concluded:

It is our strong conviction that, for a number of reasons, recovery and reuse of urban wood waste will become a more viable and attractive option in the near future:

1. *Rising costs and decreasing availability of forest-derived, primary wood for paper, fiberboard, and similar manufacturing are making alternative sources of wood fiber more attractive. Income derived from using urban waste wood, even though not enough to cover recovery costs, will help offset the costs of waste disposal. This potential income should continue to increase at least as fast as disposal costs and thus should be viewed as a buffer against rising costs.*
2. *Demand and costs for energy are rising at high rates. As the cost of energy increases relative to other costs, the option of using wood (and other organic wastes) for energy production becomes more attractive. The Energy Research and Develop-*

ment Administration has estimated that by 1985 the U.S. will have a quantity of solid waste available to produce the equivalent of 500,000 barrels of oil per day.

3. *Costs for landfilling operations and sites are increasing. In addition, space for landfilling is becoming limited, to the extent that locations are often difficult to find. As these costs rise, resource recovery and waste wood utilization become more and more attractive as a means of reducing disposal costs for all solid wastes. Comprehensive recovery programs can reduce solid waste volumes by 75 to 95 percent.*

4. *Technology for waste wood recovery seems to be in its infancy. Systems designed to produce energy, separate usable wood and other resources, and involve the public have been tested in only a few locations for relatively short periods of time; thus, some of the negative conclusions are perhaps premature. There are too many success stories and too many changing conditions to conclude that wood recovery is not feasible. We should keep in mind that there are many objectives involved. Among these are: environmental protection through reduction of solid waste; resource conservation through reuse; and partial cost recovery. If we look at only one of these, waste wood recovery can be viewed as a failure. If we consider all simultaneously, acceptable and sufficient returns will be realized.*

As suggested by the foregoing, fuelwood plays a major role in urban wood utilization, ranging from highly sophisticated systems for municipal, industrial, and institutional energy production to home heating. The energy potential for wood is significant, with one ton of dry wood producing about 17 million BTUs. This is equivalent to approximately 13,000 cubic ft (390 cm) of natural gas, 0.54 tons (0.49 t) of coal, or two barrels of oil (Arola, 1976). Table 6.1 shows comparative BTU yields per cubic foot of oven dry wood for various species.

"High technology" systems for wood energy production involve mass processing, usually chipping, hogging, or pelletizing, and may derive energy through pyrolysis or gasification. Adoption of such systems has not been as widespread as predicted because of high interest rates and the scarcity of public funds for construction and operation. And most of those currently in operation are either not using urban tree debris, or are using relatively small amounts of these materials in their total mix. The inhibiting factors as listed by Geiger are very real. The wood energy electricity generating plant at Burlington, Vermont, for example, uses only a small percentage of urban wood, partly because of the large total need, and also because of urban

177

TABLE 6.1
Conversion of Green Weight to Oven-Dry Weight and BTU per Cubic Foot

Species	Specific Gravity		Solid Wood in Pounds	[a]BTU/Green
	Green	Oven-Dry	O.D. Wt/Green C.F.	(C.F. × 10³)
			(0.45kg)	
Osage orange	0.76	0.84	47.4	408
Black locust	0.66	0.71	41.2	354
Hickory	0.64	0.78	39.9	343
Post oak	0.60	0.79	37.4	322
Pecan	0.60	0.76	37.4	322
Honeylocust	0.60	0.67	37.4	322
Mulberry	0.59	0.69	36.8	316
Bur oak	0.58	0.76	36.2	311
Red oak	0.56	0.74	34.9	300
Sugar maple	0.56	0.67	34.9	300
Green ash	0.53	0.65	33.1	285
Black walnut	0.51	0.56	31.8	273
Kentucky coffeetree	0.50	0.57	31.2	268
Hackberry	0.49	0.56	30.6	263
Red elm	0.48	0.57	29.9	257
White elm	0.46	0.55	28.7	247
Sycamore	0.46	0.54	28.7	247
Redcedar	0.44	0.49	27.5	236
Soft maple	0.44	0.52	27.5	236
Catalpa	0.38	0.42	23.7	204
Cottonwood	0.37	0.43	23.1	199
Red alder	0.37	0.43	23.1	199
Willow	0.36	0.41	22.5	194

[a]Heat yield calculated by reduction of green (fresh) volume to 0% moisture, then taking weight □ 8600 BTU to obtain yield/cubic foot of green material.

Source: Gary G. Naughton, "University of Kansas Energy Forest Third Interim Report," Kansas State University, June, 1976 (unpublished)

wood's inherent problems. Other examples are: Grenco, Inc., in Portland, Oregon, where demolition debris, waste wood from wood processing, and other urban wood are used as boiler fuel in hardboard manufacturing; Woodex, Inc., of Brownsville, Oregon, has a pelletizing process for wood and agricultural waste; the city of Ames, Iowa has burned urban waste, including wood, for energy for a number of years; and the Georgia Forestry Commission has developed a prototype wood chip fired gasification system for institutional and other adapted use.

By far the biggest use of urban forest wood, however, is for conventional fuelwood for home heating. The pervasiveness of fireplaces, heating stoves, and wood burning furnaces in recent years has created a great demand for wood removed in urban forestry operations. This demand has resulted in a number of systems and methods for making wood available to consumers. The simplest method, and generally more adaptable to smaller cities, is to leave the wood at the site—streetsides or other accessible spots—and it disappears. Other methods are municipally operated wood concentration yards where wood may be processed and/or purchased by consumers, or deliveries made by contract.

The removal of trees from the urban forest differs substantially from harvesting in conventional forestry operations. Removal is generally sporadic, with individual trees coming from a wide area. Except for trees removed in land clearing for construction, most trees are dead when cut (except in managed "woodlot" situations within the urban forest). Individual trees frequently have to be "taken down" from the top because of buildings, utilities, and other obstructions. All of these factors greatly increase the cost of the materials removed. However, these costs have to be borne anyway, either by the city or other owners, and the wood is, in that sense, free. Thus, municipal forestry departments look for marketing opportunities to offset public costs, and private tree care firms seek markets for their "free" wood in order to increase profits.

Conventional products, in addition to fuelwood, include wood chips, paving and patio blocks, lumber, and sawdust. The city of Boise, Idaho operates a wood concentration yard where all of the above products are produced. They have a small sawmill that produces wood panelling, and the waiting list for the material is far ahead of production capacity. Many private tree care firms also operate such facilities.

The removal of other urban forest materials presents special problems. Stumps are usually ground well enough beneath the soil surface and backfilled with soil. Powerful machines for this purpose have come into common use within the past two decades. Collection and disposal of leaves can be troublesome and expensive. In recent years, rapid advances have been made in collection and handling systems. Portable shredder-compactors with vacuums are not in use. The city of Webster Groves, Missouri operates a leaf compost center where leaves are shredded and composted for use in various city planting projects. Compost material is also placed in convenient pickup locations in parks and is eagerly sought by city residents.

BIBLIOGRAPHY

Arola, Roger A., data presented at seminar, "Use of wood Residues for Fuel," June 1976, Madison, Wisconsin.

Carr, Wayne F., and J.N. McGovern, "Report on Urban Wood Waste in the United States of America," prepared for USDA Forest Service, Forest Products Laboratory, Madison, Wisconsin, unpublished.

Cannon, William N., Jr., and David P. Worley, *Dutch Elm Disease Control: Performance and Costs,* Forest Service, USDA, Research paper NE–345, Upper Darby, Pa., 1976.

Cordell, H.K., and T.W. Clements, "Urban Waste Wood: a National Perspective," Urban Waste Wood Utilization, in *Proceedings of a Conference on Alternatives to Urban Waste Wood Disposal,* USDA Forest Service General Technical Report SE–16, pp. 11–12, 1979.

Geiger, James R., *Utilizing Urban Tree Debris: An Analysis of Alternatives for Chicago, Illinois,* reprint of a cooperative effort between the USDA Forest Service, S@PF, Northeastern Area, and the Chicago Bureau of Forestry, Parkways and Beautification, pp. vii–xvii, undated.

Hoefer, P., "Negotiating Successful Contracts and Agreements," in *Proceedings of Second National Urban Forestry Conference,* American Forestry Association, Washington, D.C., 1983.

Mock, D.E., "Integrated Pest Management," fact sheet, AF–95, Cooperative Extension Service, Kansas State University, Manhattan, Kansas, 1982.

Nighswonger, J.J., "Writing Specifications for Landscape Plantings," Kansas Community Forestry Program Fact Sheet, Kansas State University, Manhattan, Kansas, 1981.

Olkowski, W. and H. Olkowski, "Establishing an Integrated Pest Control Program for Street Trees," *Journal of Arboriculture,* 1:167–172, 1975.

Plinkerton, C.J., "Contracting of Tree Work and Cost Accounting," *International Shade Tree Conference Proceedings,* Western Chapter, 44:284–293, 1968.

Radtke, K., "People, Homes, and Wildfires," in *Proceedings of the National Urban Forestry Conference,* Volume I, ESF Publication 80–003, SUNY, Syracuse, N.Y., 1978.

Sacksteder, C.J., and H.D. Gerhold, *A Guide to Urban Tree Inventory Systems,* Research Paper No. 43, The Pennsylvania State University, University Park, Pa., p. 1, 1979.

Shaw, D.C., "Superintendent's Annual Report," Monmouth County Shade Tree Commission, Freehold, N.J., 1981.

Ziesemer, D.A., "Determining Needs for Street Tree Inventories," *Journal of Arboriculture,* pp. 208–213, September 1978.

SELECTED READINGS

Brown, George E., *The Pruning of Trees, Shrubs and Conifers,* Faber and Faber, London, 1972.

Christopher, Everett P., *The Pruning Manual,* MacMillan, New York, 1964.

Fenska, Richard R., *The Complete Modern Tree Experts Manual,* Dodd Mead, New York, 1964.

Harris, R.W., *Arboriculture: Care of Trees, Shrubs, and Vines in the Landscape,* Prentice-Hall, Inc., Englewood Cliffs, New Jersey, 1983.

Pirone, P.P., *Tree Maintenance,* Fourth Edition, Oxford University Press, New York, 1972.

7
MONETARY
VALUES OF
THE URBAN FOREST

In Chapter 4 we described the benefits of trees in urban areas. However, it is one thing to list these benefits and quite another to assign monetary values to them. In many cases the values are intangible and thus difficult to determine monetarily. Yet urban trees do have value both from an individual and community standpoint. Kielbaso (1971) in an article titled "Economic Values of Trees in the Urban Locale" listed eight aspects in deriving tree values: (1) alternative values; (2) city assets; (3) maintenance values; (4) timber values; (5) property values; (6) legal values; (7) evaluation formulas; and (8) replacement costs.

Alternative Values

Certainly trees must have some value or society would not spend money planting and maintaining them. Instead these funds would be invested in some other manner, for example, bank, savings and loan, stocks, and bonds. These are what are termed alternative values. Specifically, the planting and maintenance of trees in urban areas indicate the value placed on them. For example, Washington Square, a tree-covered park in Philadelphia, has an assigned property value equal to the value of the 20-story Penn Mutual Building adjacent to it (Kielbaso, 1971). Central Park in New York City is another similar example. It does not produce revenue and it costs a considerable amount to maintain the plants within its boundaries, yet New Yorkers have decided that the park is worth the cost. Even despite the recent economic problems in New York, and other cities with similar park resources, this attitude is probably unchanged.

City Assets

Given the foregoing, trees must represent an asset to a city just as streets, sewers, water lines, schools, and hospitals do. The following combination of tables from Dressel (1963) and Kielbaso (1971) shows the value of trees in urban areas as determined by a survey of city foresters (Table 7.1).

As shown in Table 7.1, cities vary in assigning value to trees. If the average table value of $187 per tree is applied to the estimated 32.25 million street trees in the United States (Tilford, 1957), it gives a total value of around $6 billion.

More recent estimates of per tree values range at between $544 and $1,714 (ICMA, 1982). Using the 32.25 million figure for street trees, one can estimate total value at between $17 and $54 billion!

The main purpose for establishing the value of the street trees is to

TABLE 7.1

Street Trees as City Assets
(Combined from Dressel (1963)
and Kielbaso (1971))

City	Date of Survey	Estimated Tree Value	Number of Street Trees	Dollar Value per Tree
Dearborn, Mich.	1958	$15,000,000	100,000	150
Detroit, Mich.	1970	60,000,000	300,000	200
East Orange, N. J.	1963	4,365,320	27,000	175
Newark, N. J.	1949	6,000,000	60,000	100
Iowa City, Iowa	1960	4,193,560	8,594	488
Greenwich, Conn.	1932	12,000,000	60,000	200
Palo Alto, Calif.	1956	6,000,000	30,000	200
Washington, D. C.	1965	25,000,000	350,000	71
Lincoln, Neb.	1966	6,000,000	60,000	100
		Average		$187

confirm to the city government and the taxpayers that trees represent a considerable investment that justifies the funding of a continuing management program.

Table 7.2 reports the results of a 1982 survey regarding the number of street trees in representative size cities.

TABLE 7.2

Mean and Median Numbers of Street Trees in Cities

Population Group	No. of Cities Reporting	Mean	Median
Total, all cities	344	26,818	11,324
Over 1,000,000	2	455,000	250,000
500,000–1,000,000	8	162,037	103,888
250,000– 499,999	13	87,038	50,000
100,000– 249,999	36	50,557	39,838
50,000– 99,999	69	27,160	20,000
25,000– 49,999	93	13,128	10,000
10,000– 24,999	113	8,432	4,000
5,000– 9,999	5	1,903	985
2,500– 4,999	5	2,070	150

TABLE 7.3

Street Tree Budget Information for
Selected Cities (from Kielbaso, 1971)

City	Number of Trees	Budget	Dollars/ Tree	City Population	Dollar/ Capita	Date Year
East Orange, N. J.	27,000	$ 102,881	3.80	90,000	1.14	1965
Richmond, Va.	250,000	168,000	0.67	219,000	0.76	1965
Wooster, Ohio	2,595	3,800	1.49	18,000	0.21	1965
Washington, D. C.	350,000	980,000	2.80	764,000	1.28	1965
Minneapolis, Minn.	350,000	1,049,000	3.00	483,000	2.17	1969
	50,000	50,000	1.00	41,000	1.22	1956
Palo Alto, Calif.	240,000 app.	649,162	2.70 app.	344,000	1.88 app.	1969
Long Beach, Calif.	33,000	324,000	9.80	108,000	3.00	1970
Lansing, Mich.	300,000	2,200,000	7.30	1,670,000	1.30	1970
Detroit, Mich.						
Midwest Average[a]			2¢—2.47—14.82		1¢—0.86—6.16	

[a]From Hatcher, 1965.

Maintenance Values

Maintenance costs are another way of determining street tree values, for example, planting, pruning, feeding, irrigation, disease, and insect control. A 1965 survey reports per capita maintenance expenditures for street trees, ranging from less than one cent per tree to $6.16, with an average of $2.47 per tree (Table 7.3) (Hatcher, 1965). These expenditures were divided as follows: planting, 14 percent; spraying, 14 percent; pruning, 32 percent; removal, 32 percent; and preventive maintenance, 8 percent. If this money were invested in a bank instead of a tree program and an annuity principle applied, the value of the trees could be estimated at different periods as shown in Table 7.4 (Kielbaso, 1971). In 1980, the national mean average expenditure per street tree was $10.78 (median $6.28) (ICMA, 1982). Thus, using these more recent figures and current interest rates, annuity values as listed in Table 7.3 would be considerably greater.

The values shown in the tables represent normal tree care. Such values can increase tremendously during insect or disease outbreaks. For example, Lincoln, Nebraska initially appropriated $135,000 to protect its elms from Dutch elm disease at a cost of $2.25 per tree. This was followed with additional appropriations of $90,000 ($1.50 per tree) for the next five years. Similar situations were common in many cities as Dutch elm disease advanced westward.

Tree care budgets also indicate tree values. New York City spends $750,000 annually to plant 5000 trees ($150/tree) (Schiff, 1969). In 1955, Richfield Oil Company spent $20,000 to plant 13 trees in the downtown area of Los Angeles ($1525/tree). Other examples are given in Tables 7.5, 7.6, and 7.7.

Table 7.5 gives mean and median tree care budgets as a percentage of the total budget. Table 7.6 provides mean annual per capita expenditures for tree care, and Table 7.7 provides information on tree planting in various sized cities from 1978 through 1981. Tables 7.5 through 7.7 are from the ICMA 1982 report on Municipal Tree Management.

Timber Values

If we were to use forestry in its traditional sense, the urban forest value would be expressed in terms of sawtimber, veneer, and other products. It has been estimated that 100,000 board feet of saw timber could be harvested on a sustained annual basis from Boston, Massachusetts (Foster, 1965). Obviously, in the urban areas of North America the amenity values of trees make this approach difficult. It should be noted that trees are an asset that

187

TABLE 7.4

Annuity Values for 30 and 50 Years at 5% Interest
for Selected City Forestry Programs
(from Kielbaso, 1971)

City		Total Program Value			Individual Tree Values		
		Budget Information	30 Years	50 Years	Expenditures Per Year	30 Years	50 Years
East Orange, N. J.	1965	$ 102,881	$ 6,831,300	$21,502,129	$3.80	$252.32	$ 794.20
Richmond, Va.	1965	168,000	11,155,200		.67	44.49	140.43
Washington, D. C.	1965	980,000	64,972,000		2.80	185.92	585.20
Wooster, Ohio	1965	3,800	252,320		1.49	98.94	311.41
Detroit, Mich.	1970	2,200,000	145,200,000		7.30	481.80	1,525.70
Lansing, Mich.	1970	352,000	23,372,800		9.89	646.80	2,048.00
Minneapolis, Minn.	1969	1,049,000	70,313,600		3.00	199.20	627.00
Palo Alto, Calif.	1956	50,000	3,320,000		1.00	66.40	290.00
Midwest Average[a]	1965				2.47	164.01	516.23

[a]From Hatcher, 1965.

TABLE 7.5

Mean and Median Tree Care Budget as a Percentage of Total Municipal Budget

Classification	No. of cities reporting	Mean tree care budget ($)	Mean as % of total municipal budget	Median tree care budget ($)	Median as % of total municipal budget
Total, all cities	946	131,023	0.81	36,000	0.40
Population group					
Over 1,000,000	5	3,128,280	0.25	1,975,986	0.01
500,000–1,000,000	10	1,050,474	0.33	800,000	0.23
250,000–499,999	24	828,726	0.42	375,000	0.19
100,000–249,999	65	299,784	1.01	232,831	0.49
50,000–99,999	132	181,029	0.94	124,000	0.56
25,000–49,999	225	89,515	1.02	50,000	0.52
10,000–24,999	428	32,358	0.69	14,500	0.29
5,000–9,999	27	11,890	0.43	4,500	0.33
2,500–4,999	30	7,310	0.57	3,000	0.28
Geographic region					
Northeast	250	65,643	0.54	15,000	0.25
North Central	314	170,715	1.20	50,000	0.64
South	160	73,320	0.47	20,000	0.17
West	222	190,095	0.79	68,044	0.49
Metro status					
Central	203	366,640	0.74	142,985	0.32
Suburban	524	79,454	0.92	28,000	0.48
Independent	219	36,008	0.62	13,600	0.26
Form of government					
Mayor-council	326	170,146	0.66	20,000	0.25
Council-manager	538	116,689	0.88	50,000	0.49
Commission	28	123,011	0.78	50,345	0.57
Town meeting	38	29,209	0.65	11,500	0.27
Rep. town meeting	16	71,685	1.30	60,000	0.27

TABLE 7.6

Mean Annual per Capita Expenditures for Tree Care

Classification	No. of cities reporting	Mean ($)
Total, all cities	945	2.19
Population group		
Over 1,000,000	5	1.42
500,000–1,000,000	10	1.58
250,000– 499,999	24	2.42
100,000– 249,999	65	2.11
50,000– 99,999	132	2.51
25,000– 49,999	225	2.52
10,000– 24,999	427	1.98
5,000– 9,999	27	1.59
2,500– 4,999	30	2.09
Geographic region		
Northeast	250	1.31
North Central	314	3.06
South	160	1.19
West	221	2.66
Geographic division[1]		
New England	128	1.54
Mid-Atlantic	122	1.08
East North Central	209	2.88
West North Central	105	3.41
South Atlantic	86	1.51
East South Central	16	0.50
West South Central	58	0.90
Mountain	53	1.95
Pacific Coast	168	2.88
Metro status		
Central	203	1.99
Suburban	523	2.38
Independent	219	1.90

[1]*Geographic divisions: New England*—the states of Connecticut, Maine, Massachusetts, New Hampshire, Rhode Island, and Vermont; *Mid-Atlantic*—the states of New Jersey, New York, and Pennsylvania; *East North Central*—the states of Illinois, Indiana, Michigan, Ohio, and Wisconsin; *West North Central*—the states of Iowa, Kansas, Minnesota, Missouri, Nebraska, North Dakota, and South Dakota; *South Atlantic*—the states of Delaware, Florida, Georgia, Maryland, North Carolina, South Carolina, Virginia, and West Virginia, plus the District of Columbia; *East South Central*—the states of Alabama, Kentucky, Mississippi, and Tennessee; *West South Central*—the states of Arkansas, Louisiana, Oklahoma, and Texas; *Mountain*—the states of Arizona, Colorado, Idaho, Montana, Nevada, New Mexico, Utah, and Wyoming; *Pacific Coast*—the states of Alaska, California, Hawaii, Oregon, and Washington.

Number of Trees Planted in 1978 and 1979
and Number Estimated for 1980 and 1981
(From ICMA, 1982)

Classification	No. Planted—1978			No. Planted—1979			No. to Be Planted—1980			No. to Be Planned—1981		
	No. of Cities Reporting	Mean	Median	No. of Cities Reporting	Mean	Median	No. of Cities Reporting	Mean	Median	No. of Cities Reporting	Mean	Median
Total, all cities	651	517	165	730	503	150	680	519	160	542	574	190
Population group												
Over 1,000,000	4	6,852	2,000	4	7,755	3,500	4	6,213	3,500	3	7,043	130
500,000–1,000,000	8	2,865	2,500	8	4,083	3,000	9	3,483	2,500	7	4,071	3,000
250,000–499,999	20	3,312	2,000	20	3,206	1,700	20	2,985	1,800	20	3,883	1,600
100,000–249,999	45	1,158	791	48	1,122	556	51	1,065	500	40	1,090	500
50,000–99,999	104	578	350	114	549	311	106	531	300	80	537	350
25,000–49,999	165	307	198	190	292	150	176	325	200	143	313	200
10,000–24,999	273	196	100	306	203	89	282	232	100	222	226	100
5,000–9,999	14	141	65	17	210	40	17	109	30	15	129	40
2,500–4,999	18	102	32	23	92	30	15	149	20	12	61	35
Geographic region												
Northeast	189	201	88	205	176	80	179	188	96	138	167	100
North Central	239	699	235	262	659	234	252	660	250	202	691	250
South	80	413	150	101	435	100	92	472	130	80	724	200
West	143	677	200	162	708	200	157	699	200	122	744	200
Geographic division												
New England	95	159	75	104	135	72	91	158	80	72	131	100
Mid-Atlantic	94	262	100	101	217	90	88	220	100	66	207	90
East North Central	161	598	200	175	613	200	171	552	200	135	613	250
West North Central	78	907	250	87	753	280	81	888	300	67	849	250
South Atlantic	49	516	200	61	527	125	57	557	150	49	994	200
East South Atlantic	9	210	75	11	177	30	7	96	50	6	122	150
West South Atlantic	22	266	100	29	338	50	28	394	100	25	338	150
Mountain	23	452	115	30	626	150	32	534	85	25	596	100
Pacific Coast	120	721	228	132	727	200	125	741	200	97	782	200

tends to increase in value over time. They do, however, become a liability at death because they have to be removed. This liability could perhaps be partially offset by a sustained yield management program. This liability can be further offset by the utilization and sale of chips, firewood, and other products. However, this is not always feasible and depends on local market conditions.

Property Values

Most people consider that trees do enhance property values. However, there are those who assign a negative value as their presence means the raking of leaves and problems in lawn maintenance. Tree value is manifested by the final agreement made between buyer and seller. Is the buyer willing to pay more for a wooded lot as compared to a treeless one? Tree values are hard to assess in these situations as it is difficult to separate tree values from the rest of the property (land and improvements).

Kielbaso (1971) contacted a bank appraiser concerning tree values in subdivisions. The appraiser suggested that "off the record" he might value a lot with trees at up to $10 per front foot more than a lot without trees. However, he could not be definitive because there is much risk involved. Kielbaso (1971) suggests that it really boils down to two factors: high grade buyers (higher income bracket) and risk. He suggests that people with high incomes are more willing to pay for the amenities trees provide and banks are more willing to risk their money on such affluent individuals.

Kielbaso (1971) gave the following examples of indirect methods to ascertain the property values assigned to trees in urban areas: (1) Gould (1970) reported that property values in the northeastern United States seemed to vary with the amount of forest and open space and that the best of the suburbs had almost 100 acres of forest or open space per 1,000 people; (2) Edlin (1963) reported a situation of urban dwellers opting to preserve a 10-acre woodland adjacent to their property primarily because they feared a loss of market value if the woods were to be removed; (3) Purcell (1956) reviewed some prime property subdivisions in Toronto in which small orchard trees aided in selling the homes. One developer valued the trees at $300 each and planned his streets to utilize a maximum of them; and (4) national realty officials have adopted increased values for homes having sound tree plantings (Johnson, 1956). Kielbaso (1971) mentions also a home with a 300-year-old elm being sold. A $24,000 offer was made prior to a hurricane that destroyed the tree. The undamaged home eventually sold for $15,000 after the hurricane. Thus, it could be concluded that tree was worth $9,000 to the property value.

Dr. Brian Payne of the U.S. Forest Service has found that trees may increase property value by as much as 20 percent, with average increases of 5 to 10 percent. This translates to increases of $3,000 to $7,000 in value. The data were obtained by showing photographs of architecturally similar houses, with and without trees, to realtors and asking them to estimate selling prices (Payne, 1975). In another study of a similar nature, seven simulated combinations of amount and distribution of tree cover for a 12-acre tract of unimproved residential land were presented to professional appraisers for value estimation. Arrangements with trees were valued higher than those without trees and scattered tree arrangements were valued higher than dense cover arrangements (Payne and Strom, 1975).

As can be seen, trees do have value in real estate. However, the actual value is subject to discussion and varies from region to region with realtors, developers, and individual buyers and sellers.

Legal Values

Probably the area most fraught with difficulty in assigning the worth of trees is that of legal values, that is, court suits, income tax deductions, and insurance casualty loss claims. Recently we have seen some trends toward the viewpoint that trees should have legal standing or legal rights. Indeed there is a book by Christopher D. Stone titled "Should Trees Have Standing—Towards Legal Rights for Natural Objects" (1974). In the past, an unwillingness to attach values to trees contributed to society's general disregard of trees. This is evidenced by settlements that insurance companies award for tree damage and the general attitude of the Internal Revenue Service toward tree values. However, as evidenced by the recent environmental movement, people are more concerned about trees and their legal values. This has caused insurance claims adjusters, Internal Revenue Service agents, and municipal officials to become more aware of tree values. Below are some examples of tree appraisal situations that have been reported in the Rochester, New York area over the past several years (Micha, 1975):

- Tree damage due to vehicular accident
- Sudden tree loss due to high wind
- Presurvey of trees prior to gasline installations
- Presurvey of trees prior to underground electrical installations
- Presurvey of trees prior to sewer and water-line construction
- Overencroachment during construction
- Presurvey of trees prior to road widening

- Trees damaged or destroyed by lightning
- Timber theft
- Poor pruning practices
- High water tree loss
- Chemical injury to trees and plants
- Property survey to determine tree values prior to home construction
- Tree damage and loss due to fire

If one were to survey other cities, other situations could certainly be found, but it does point out the increasing concern of landowners for tree values. In terms of such appraisals, let us look at the normal channels in which legal values for trees are ascertained and hopefully awarded; that is, insurance companies, the Internal Revenue Service, and the appraisal process and the courts.

Insurance settlements

Insurance agencies regularly deal with casualty losses to trees. One of the first problems that enters into such a discussion is what is a casualty. As defined by the Internal Revenue Service, "a casualty is the damage, destruction or loss of property resulting from an identifiable event of a sudden, unexpected or unusual nature." Loss of death of a tree due to storm, flood, or fire is a casualty loss. But what about Dutch elm disease or other insect or disease attacks? IRS Handbook #17 (revised November 1972) states that the damage or destruction of trees, shrubs, or other plants by a fungus, disease, insects, worms, or similar pests is not a deductible casualty loss. However, a sudden, unexpected, or unusual infestation by beetles or other insects *may* result in a casualty loss.

Casualty loss statements in all insurance policies are essentially the same. The various insurance companies do not usually dictate the terms of their policies nor do the state insurance commissions. As a whole, there are about three or four insurance service organizations that write the policies for all the rest of the insurance companies in the nation (Gustin, 1975). Thus, the policies from all the companies tend to be similar and read as follows (Gustin, 1975):

> *This company shall be liable for loss to trees, shrubs, plants, and lawns (except those grown for business purposes) only when the loss is caused by fire, lightning, explosion, riot, civil commotion, malicious mischief, theft, aircraft or vehicle not operated by an occupant of the premises. The Company's liability in any one oc-*

*currence under this provision shall not exceed in the aggregate for
all such property 5 percent of the limit of liability of Coverage A
nor more than $250 on any one tree, shrub or plant including
expense incurred in removing the debris.*

Most companies do have some type of deductible clause ($50 to $100)
and thus for any one tree the actual amount collectible for a casualty could
be as low as $150. If a large tree was involved, it may well cost in excess of
$250 to simply remove the tree, let alone replace it. In addition, it does not
matter if you have a valuable rare tree or if you live in California, Kansas,
or New York, the limit is still the same. Insurance companies indicate that
the number of claims for tree losses are few (Gustin, 1975). The limit of
$250 was established in 1954 and has remained the same since. Any policy
change has to involve the Board of Directors of the Insurance Service Organization and they meet only once every six years (Gustin, 1975). The
International Society of Arboriculture and the American Society of Consulting Arborists are attempting to persuade the insurance industry to increase liability coverage under the casualty clause to $500.

Internal revenue service

Claiming casualty losses as income tax deductions is a much more complex
procedure with many variables. Among these are: (1) rulings of a similar
nature involving trees and shrubs that may be referred to by IRS personnel
when reviewing a casualty loss claim; (2) the attitudes of reviewing officials
toward trees and their legal standing or value; and (3) how closely the
individual return is reviewed.

The IRS publication, *Tax Information on Disasters, Casualty Losses and
Thefts*, makes the following points in reference to trees and shrubs:

*Damage to trees and shrubs. If ornamental trees and shrubs on
residential property are damaged or destroyed by a casualty, you
may have a deductible casualty loss. To claim a deduction, you
must establish that there has been a decrease in the total value of
the real estate. A different rule applies in cases of such damage to
property used for business purposes.*

Proof of Loss:

*You must be able to prove that you actually sustained a casualty
or theft loss and the amount of the loss that is deductible. You*

195

should be prepared to show:

 1. *The nature of the casualty and when it occurred (in the case of theft, when it was discovered).*

 2. *That the loss was the direct result of the casualty (or in the case of theft, that the property was actually stolen).*

 3. *That you were the owner of the property.*

 4. *The cost or other basis of the property, evidenced by purchase contract or deed, for example, (improvements should be supported by checks, receipts, etc.).*

 5. *Depreciation allowed or allowable, if any.*

 6. *Value before and after the casualty, for example, decreases in fair market value.*

 7. *The amount of insurance or other compensation received or recoverable, including the value of repairs, restoration, and cleanup provided without cost by disaster relief agencies.*

If the preceding material is read carefully, three important factors stand out when computing the amount of a casualty loss for IRS purposes. These are:

1. The *decrease* in *fair market value* of the property as a result of the casualty.
2. The *adjusted basis* of the property.
3. The *amount of insurance or other compensation* you received or expected to receive as a result of the casualty.

In general, the amount of loss is the decrease in fair market value, limited to the adjusted basis, and reduced by any insurance or other reimbursement.

The problem of insurance compensation is straightforward. However, what is meant by *decrease in the fair market value* and *adjusted basis of property?*

Decrease in the Fair Market Value: This is the difference between the value of the property immediately before and immediately after the casualty. It is not necessarily the cost of replacing or repairing the property and most certainly does not reflect any sentimental value. The loss has to be based on intrinsic value, apart from any sentimental or replacement value. The difference between the fair market value of the property immediately before and immediately after the casualty should be established by contacting an experienced and reliable appraiser. Any fees involved in this appraisal may be deducted as an expense of determining the tax liability but not as a part of the casualty loss deduction.

Adjusted Basis of Property:
Most property is acquired by purchase. The amount paid for it is termed its "basis." However, if the property has been subjected to depreciation, previously casualty losses, or other recovery adjustments, the "basis" of the property must be reduced to reflect these amounts. Conversely, any expenditures for capital improvements should be added to the "basis." The end result is what is termed the "adjusted basis."

IRS $100 Limitation:
In the computation of deductions for personal casualty losses the IRS requires that $100 be deducted from the final total loss value. For example, if the total loss is $2,100 the actual amount you are allowed to claim is not $2,100 but $2,100 minus the $100 limitation, or $2,000.

Below are two examples of casualty loss deductions. One involves personal casualty loss deduction, the other a business casualty loss deduction. Both involve landscape plant considerations.

Personal Casualty Loss Deduction:
Different rules apply when determining the amount of personal casualty loss deduction in the case of real property (house and landscaping) and personal property (furniture, car, etc.).

In determining a loss to real property, all improvements, such as buildings and ornamental trees are considered an integral part of the realty, and a single casualty loss for the entire property is determined. The amount of loss deductible is computed by comparing the decrease in the fair market value of the entire property with its adjusted basis. The loss for tax purposes is the lesser of these amounts. From this amount you must subtract any insurance reimbursements, and finally the $100 limitation must also be deducted. For example:

You purchased a house several years ago paying $3,000 for the land and $15,000 for the building. You also paid $1,000 for landscaping. In June the residence was completely destroyed by fire. Competent appraisers determined that a fair market value immediately before the fire of $22,000 but only $5,000 after the fire. Shortly after the fire, the insurance company paid

197

you $15,000 for the loss of your residence. Your casualty loss would be computed as follows (IRS, 1976, pages 4 and 5):

1.	Value of the total property before the fire	$22,000.00
2.	Value of the total property after the fire	5,000.00
3.	Decrease in value of total property	$17,000.00
4.	Adjusted basis of entire property (land, building, and landscaping)	$19,000.00
5.	Loss sustained (lesser of 3 or 4)	$17,000.00
6.	Less insurance recovery	15,000.00
7.	Casualty loss before application of $100 limitation	2,000.00
8.	Less $100 limitation	100.00
9.	Casualty loss deduction	$ 1,900.00

However, what about damages to trees and shrubs? For example, you have paid $3,000 for land and $30,000 for a new home. In addition you have spent $1,250 landscaping, which includes $1,000 for a 6-in. caliper, 25-ft tall blue spruce. A windstorm completely destroyed the blue spruce tree but left the house and remaining landscape plants unscathed. How would this casualty loss be calculated? The IRS has stated that trees and plants are considered an integral part of the property and no determination of fair market value of the house, land, and trees can be made separately. This presents a dilemma. However, there is a bright spot—read carefully the following excerpt out of the IRS booklet (page 5) on casualty losses.

> If a storm damages trees or shrubs on property held for personal use (real property), the loss is the actual decrease in the value of the property as a whole. The cost of repairing the damage to the property may, in any particular case, serve as evidence of the overall loss to the property. However, the loss may not exceed the amount necessary to restore the property to its value just before the casualty.

Thus, replacement costs are acceptable as evidence of decrease in value. For example, a qualified appraiser valued the total property at $39,000 before the storm. This example is assuming some appreciation, which has been the rule for most homes in recent years. A nurseryman estimates replacement cost at $1,500 for the blue spruces.* Your casualty loss deduc-

*This evaluation could also be done by an urban forester, horticulturist, or arborist.

tion would be as follows:

1.	Value of the real property before the storm (cost of land, building, and landscaping)	$39,000.00
2.	Value of the total property after the storm, $39,000 − $1,500	$37,500.00
3.	Decrease in the value of the total property	$ 1,500.00
4.	Adjusted basis of the entire property $3,000 + $30,000 + $1,250	$34,250.00
5.	Loss sustained (lesser of 3 or 4)	$ 1,500.00
6.	Less insurance recovery ($250 per tree)	250.00
7.	Total loss	$ 1,250.00
8.	Less $100 limitation	100.00
9.	Casualty loss deduction	$ 1,150.00

As we will discuss later, the problem of shrub and tree losses focuses mostly on how the loss is arrived at, that is, replacement costs versus individual shade-tree evaluation formulas.

Business Casualty Loss.　Fortunately for businesses, the IRS is a little more lenient in claiming losses to landscape plants. With business property, the fair market value involves the individual items of property destroyed. Thus, if a business property loses ornamental plants or trees, the amount of the deduction is by reference to the volume of the plants lost and does not involve any change in the value of the entire property (Gustin, 1974). As stated by the IRS manual (page 5):

> A casualty loss to the business property or to property held for the production of income is determined by reference to the identifiable property damage. Thus, if damage occurs to a building and trees, the decrease in value is measured by taking each into account separately, with separate losses being determined for each.

For example, three years ago you purchased a house for rental purposes paying $3,000 for the land and $15,000 for the building. You also paid $1,000 for landscaping. For those years you were allowed total depreciation of $1,125 for the building and $150 for the landscaping. In June, the house was completely destroyed by fire and severe damage was inflicted on the landscaping. Competent appraisers determined that the trees and shrubs were worth

$1,500 before the fire but only $1,000 after the fire. The trees and shrubs were not covered by insurance but the house was insured for its fair market value. Shortly after the fire, the insurance company paid you $17,500 in full settlement of its liability.

The gain or loss from the fire is determined as follows (IRS, 1976, p. 5):

		Building	Trees and Shrubs
1.	Value before fire		$1,500
2.	Value after fire		1,000
3.	Decrease in value		$ 500
4.	Adjusted basis	$13,875	$ 850
5.	Loss for trees and shrub decrease in value		$ 500
6.	Less insurance recovery	$17,500	None
7.	Gain from fire insurance proceeds on building	$ 3,625	
8.	Loss from fire damage to trees and shrubs		$ 500

As shown thus far in this section on legal values, it is not impossible to collect insurance payments or deduct tax losses for the casualty loss of trees and shrubs. However, it is important to understand the regulations in order to do so competently. With the material just discussed as source material and with their professional background, urban foresters are qualified to make the necessary appraisals. With time, insurance and IRS regulations will change and all should be reviewed and kept up to date for accurate appraisal.

Litigation and condemnation

Often foresters and arborists are asked to estimate tree values in condemnations and civil suits. Condemnation, according to the IRS, is "the exercise of a legal power of the Federal Government, a State Government, or political subdivision, to take privately owned property for any public use, upon an award and payment of a reasonable price for the property."

These usually involve city improvement projects, for example, new streets, parks, sewer lines, and water lines. They may also involve commercial utility firms rights-of-way (e.g., power and light companies and natural gas companies). Land having trees may be condemned, thus forcing an evaluation of the timber in addition to the land. In other cases, this may involve only the value of the trees removed. The evaluation procedure used depends on the use of the property. In some cases, forest timber evaluation procedures are appropriate. Where esthetic benefits of trees in urban areas or around rural homes are of paramount importance, the International Society of Arboriculture Shade Tree Evaluation Formula is used. The use of this formula is discussed later in this chapter.

In recent years the number of civil suits involving trees has increased substantially. This reflects the increasing concern for the legal values of trees and attempts to attach monetary values to what were previously considered intangible benefits. Such suits most often involve construction activities, herbicide drift, vandalism, and improper tree maintenance activities. A common factor in all of these suits is an appraisal of damage, which requires the professional expertise of a forester or an arborist using timber or shade tree evaluation procedures. Many suits end up in court, require additional testimony from the professional involved, and litigation may be prolonged for years. Kielbaso (1971) and Peters (1971) have presented information regarding litigation, IRS, and insurance settlements. Kielbaso (1971) in regard to litigation has stated:

> *Through time and many appealed cases, the following principles appear noteworthy:*
> 1. *Most cases involving tree values are determined by measure of property value before and after tortious acts (this is most common, but not prescribed).*
> 2. *Under appropriate circumstances, justice may dictate that the decision be based on the value of the tree or the reasonable cost of a reasonable restoration:*
>   a. *The owner of an estate is entitled to have it in such condition as he wants.*
>   b. *If an owner holds for a purpose, be it his own use or sale, he is entitled to compensation although the damage does not impair general market value.*
>   c. *But an ornamental shade tree, upon land available for dwelling-houses, has a very different relation to the land, and may give it a special value.*

Kielbaso (1971) in citing cases under each of the above categories made the observation that:

> *Perhaps the point to note is that in the more recent cases the courts have been awarding damages for trees other than the property market-value reduction. They have even approached esthetics and even sentimental values, although this terminology is carefully avoided.*

201

For the professional doing the evaluation for litigation the process more often than not is more complex than applying a simple formula. Often other factors come into focus such as woodland values versus residential values, simple shade tree evaluation versus replacement costs, cleanup costs, and site restoration costs.

Evaluation Formulas

As mentioned earlier the assigning of dollar values to shade and ornamental trees is important for a variety of reasons: determining city tree assets for budgetary purposes; insurance compensation for casualty losses; income tax deductions; condemnation; and litigation. The use of established shade tree evaluation formulas is very common and an accepted methodology in all of these except for income tax deductions. However, replacement costs can be applied in the case of IRS casualty loss deductions.

Shade tree evaluation formulas have evolved over the past century. One of the first involved an arbitrary system of valuation of $5 to $150 per tree. It was used to ascertain damage caused by horsedrawn carriages. In 1901, a professor at the University of Michigan devised a formula for shade tree valuation using a set figure of $15 plus compound interest at four percent for 25 years. This was known as the Roth Method. Sometime later a circumference method was devised that assigned an arbitrary value of $5 per circumference inch. This was followed by a diameter inch on a sliding scale and later a basal square foot and basal square inch method set values at $75 per square foot and 75¢ per square inch. The earliest generally accepted method was developed in the early 1900s by Dr. George Stone of the Massachusetts Agricultural College. His formula used as critical factors the size, location, and condition of trees. Dr. Stone later worked with Dr. E.P. Felt of the Bartlett Research Laboratories in an attempt to improve evaluation procedures. The result was the Felt method, which assigned a value of $1 per square inch of basal area at diameter breast height (dbh) and allowed for species, condition, location, and residential land values. The Felt formula was widely accepted and used by those in arboriculture for many years. In 1969, the Felt method was revised by O.W. Spicer for use by the Bartlett Tree Experts Co. (Spicer, 1969). This is known as the Felt-Spicer Formula. It takes into account the following factors for computing tree values: size, species, basic value, location, condition, fluctuating land values, and the changing purchasing power of the dollar (inflation). The basic value is computed by determining the diameter of the tree and using the formula [$R^2 \times$ (cost of living index) = basic value in dollars] when R^2 is the radius of the tree at dbh squared. The base value is then adjusted up or down by the application of a species factor, condition factor, location factor, and an assessed property evaluation factor. The primary arguments against this method are that it is somewhat discriminatory to use assessed property values and that it also requires the additional expertise of a land assessor. Most foresters and arborists feel that a tree should have a value of its own separate from the value of the property involved.

Perhaps the most widely accepted method for shade tree evaluation is the one developed jointly by the International Shade Tree Conference (ISTC) and National Arborists Association (NAA). The method was first developed in 1951 and was accepted and published in 1957. Since that date, there have been five revisions of the publication: 1965, 1970, 1975, 1979, and 1983. Basically the ISTC Formula as originally devised was as follows:

Value of tree = basic value × species factor × condition factor

The basic value was determined by measuring the diameter of the stem at a point 4.5 (1.37 m) ft above the ground (dbh). This was then converted to square inches and multiplied by $5. Since 1957 the dollar value has been revised five times to account for inflation, and it now stands at $22 per square inch.

Species factor was based on local experience with various species in certain well-recognized geographic regions. It was expressed in five percentages classes of 100, 80, 60, 40, and 20.

Condition factor was also expressed in five classes: 100, 80, 60, 40, and 20 percent. In this matter, considerations of age, location, disease, and so on were to be weighed.

Over the years, the ISTC formula has been revised and refinements added. The previous formula was not without fault. It presented no procedure for evaluating shrubs, underestimated values for smaller trees, overestimated values for large trees, and did not incorporate a location factor. The current formula stresses the professional aspects of shade tree evaluation. It also incorporates guidelines to allow for replacement values to be used for shrubs and for small trees under 12 in. (30.48) in diameter. This is normally done by consulting with nurseries that can supply the appropriate stock of a given species in the size desired. It also considers such factors as availability, area problems, removal costs of casualty trees, and guarantees. Normally, the value exceeds the $22 per square inch category used in larger trees. The actual calculation for trees and shrubs under 12 in. in diameter is as follows:

Tree value = basic replacement cost × species factor × condition factor × location factor

203

The basic value is computed using tree caliper to determine the basal area in square inches. This is done by applying the American Standards for Nursery Practice in caliper measurements: for trees up to 4 in. in caliper the basal area is computed using tree caliper measurements at 6 in. above the ground and 12 in. above the ground for those over 4 in. in caliper; and a location factor has been added in five classes of 100, 80, 60, 40, and 20 percent. This precent class is judged from architectural, engineering, climate, and esthitic benefits that take into account such funtional values as shade, screening, sound, and climate amelioration. It can also be extended for use on trees in park settings.

Trees that range from 13 to 40 in. (33.02 to 101.6 cm) are evaluated by the following formula:

Tree value = basic value × species factor × condition factor × location factor

Size in these larger trees is based upon cross sectional area in square inches at diameter breast height or 4.5 ft (30.48 cm) above ground level. Another revision in trees of the 13 to 40 in. (33.02 to 101.6 cm) category involves evaluation of multiple stem specimen trees.

Evaluation of shade trees above 40 in. (101.6 cm) in caliper is left to the judgment of the professional involved. Specifically this is designed to eliminate the problem of the excessively high values that such trees generate by strict application of the formula. It allows for the professional judgment of foresters, arborists, and horticulturists on an individual case basis, using the knowledge and experience that qualifies only them to make such determinations.

It is hoped that with the revisions noted, the new formula of the ISTC (now the International Society of Arboriculture) will become a standard method to be used in the computing casualty losses for not only the insurance industry and the courts but the Internal Revenue Service as well.

BIBLIOGRAPHY

Chadwick, L.C., "ASCA Recommendations for Modification of the ISTC Shade Tree Evaluation Formula," *Journal of Arboriculture,* 1(2):35–38, 1975.

Dressel, K., "Street and Park Evaluation," in *Proceedings of Midwest Shade Tree Conference,* 18:105–112, 1963.

Edlin, H., "Amenity Values in British Forestry," *Forestry,* 36:65–89, 1963.

Felt, E.P., *Our Shade Trees,* Orange Judd Publishing Co., 1942.

Foster, C., "Forestry in Megalopolis," *Proceedings of Society of American Foresters Meeting*, 65–67, 1965.

Gould, E., "Values, Trees and the Urban Realm," *Symposium on Trees and Forests in an Urbanizing Environment*, University of Massachusetts, Amherst, pp. 79–91, 1970.

Gustin, Guy, Jr., "Viewpoint of IRS and Insurance Agencies on Shade Tree Values," *Journal of Arboriculture*, 1(3):58–60, 1975.

Hatcher, O., "Scope of Shade Tree Care in Region 5 of the International Shade Tree Conference," *Proceedings of International Shade Tree Conference*, 41:114–119, 1965.

International City Management Association, "Urban Data Service Report—Municipal Tree Management," Vol. 14, No. 1, January 1982.

Internal Revenue Service, *Tax Information on Disasters, Casualty Losses, and Thefts*, Publication 547, 1976.

International Society of Arboriculture, *A Guide to the Professional Evaluation of Landscape Trees, Specimen Shrubs and Evergreens*, 1975.

Johnson, I., "Planning and Implementing a Municipal Street Tree Program," *Proceedings of International Shade Tree Conference*, 32:219–229, 1956.

Kielbaso, J. James, "Economic Values of Trees in the urban Locale," *Symposium on the Role of Trees in the South's Urban Environment*, pp. 82–94, 1971.

Micha, Frederick, R., "Trees Should Have Standing," *Journal of Arboriculture*, 1(4):78–80, 1975.

Payne, Brian R., "Trees Could Make A Difference In The Selling of Your Home," Northeastern Forest Experiment Station, Forest Science Photo Story No. 26, 1975.

Payne, Brian R. and Steven Strom, "The Contribution of Trees to the Appraised Value of Unimproved Residential Land," *Valuation*, 22(2):36–45, 1975.

Peters, Lewis C., "Shade and Ornamental Tree Evaluation," *Journal of Forestry*, 69(7):411–413, 1971.

Purcell, C., "The Realty Value of Trees," in *Proceedings of International Shade Tree Conference*, 32:128–135, 1956.

Schiff, C., "What Is Happening in Municipal Arboriculture in New York," in *Proceedings of the International Shade Tree Conference*, 45:121–217, 1969.

Spicer, O.W., *Appraising Shade and Ornamental Trees*, Bartlett Tree Experts, Stanford, Conn., 12 p., 1969.

8
MUNICIPAL
FORESTRY
ADMINISTRATION

How well the urban forest is managed depends on the attitudes, knowledge, and financial resources of its various owners. The degree of management by municipal forestry organizations is largely an expression of the priority given by city officials. This priority is in turn largely a manifestation of citizen concern. Successful management of the urban forest generally requires: (1) a positive political and policy environment; (2) strong administration; (3) effective legislation; (4) efficient organization; (5) sound planning; and (6) adequate funding. Of these, a positive political and policy environment is fundamental, as it is the basis for the others.

Political and Policy Environment

As suggested above, the political environment of municipal forestry programs is a result of citizen concern. It can be positive or negative in generally direct proportion to the voice of the citizens. As will be discussed in Chapter 9, this environment can be manipulated by public relations. Johnson (1982) states that "the policy environment around urban forestry programs can be enhanced, while maintaining an autonomous nature, by improving the decision-making and management functions. High standards should be held for goal establishment, efficient operations, and evaluation procedures." He then goes on to suggest a key element based on a study of urban forestry programs in 12 U.S. cities:

> A key element for strong urban forestry programs appears to be the existence (or at least the perception) of a "crisis" in the urban forest, such as the Dutch elm disease in some cities. When a crisis is believed to exist, the mobilization of funds, community support, assistance, and greater commitment are more frequently stimulated than when the goal of the program is maintenance. Therefore, it is essential that urban forestry programs improve their capabilities to forecast and identify deteriorating conditions so that action can be taken to deal with nascent crisis before they reach the stage where little can be done, or when effective actions are cost prohibitive.

The political environment is made up of attitudes and actions of the citizenry, and the attitudes and actions (and reactions) of elected city officials. The policy environment is a product of this, and as such is the foundation upon which programs are built. Bartenstein (1980) cites four policy goals upon which the city of Dayton, Ohio is seeking to develop a comprehensive strategy for urban forestry activity: (1) economic vitality; (2) neighborhood vitality; (3) maintenance of the city's unique character; and (4) urban conservation. He observes accurately also that:

> *Before an urban forest management strategy can be complete, it must include mechanisms for comparing the efficiency of forest strategies to other methods for enhancing a city's economic and neighborhood vitality, unique character and physical environment. . . . There must also be consideration given to the costs an urban forest imposes on the local government, its citizens and other urban services.*

Administrative Structure

Most municipal forestry programs are administered by commissions, boards, or other authorities. These generally take three forms: (1) advisory; (2) policymaking; and (3) operational.

Advisory boards or commissions have no direct authority or responsibility for policy or administration. Their duties are to study and investigate and give counsel to the city administration. They have no authority to expend funds or to conduct actual operations. Figure 8.1 shows the relationship of such boards to the city government.

Policymaking boards are semi-independent with policymaking and administrative authority for municipal forest management (Figure 8.2). Legislative authority varies, but they generally have responsibility for program planning and implementation. They are responsible for program budgets, but are answerable to higher administrative authority in program conduct and expenditure of funds. Most of the small towns involved in the Kansas Community Forestry Program operate with such boards. This program is discussed in detail in Chapter 9.

Operational commissions are independent of municipal administrative structures. They have full responsibility for policy formulation and program implementation. They are clearly distinguished from the other types by independent financing based on tax levies. Fund disbursement is the clear responsibility of such commissions and cannot be interfered with. Decision making is not passed to another body. Such commissions are similar to school boards in legislative authority and responsibility.

Operational responsibilities of commissions vary depending on organi-

209

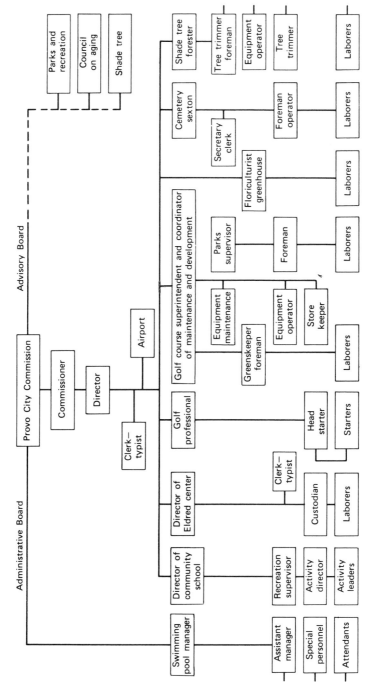

Figure 8.1 An organizational chart of a park and recreation department having an advisory shade tree board.

zational structures. Chadwick has listed some possible combinations (Chadwick, 1972):

1. A municipal street tree commission with jurisdiction over street, boulevard, and highway trees only.
2. A municipal shade tree commission with jurisdiction over all trees on streets, parkways, and in parks and public places.
3. A municipal park and/or park and forestry commission that is similar to the two commissions just mentioned but encompasses other park activities beyond jurisdiction over shade or ornamental trees.
4. A municipal park and recreation commission that combines parks and forestry.

However structured and organized, municipal forestry boards or commissions should have clearly defined responsibilities and should be as free from political interference as possible. It is very important that capable people serve. They should be public-minded citizens, dynamic, and able to make and carry out decisions.

There must be ready access to technical and legal expertise, either through the qualifications of individual board members or through available outside experts. In cities with professional staffs, technical forestry input is from the city forester. In other cases, it may be from consultants or state or federal agency professionals. Legal expertise is most often from city attorneys. Often an attorney is included as an ex officio member of the board.

Most municipal forestry functions are within or attached to park and recreation departments, or public works or similar service departments. Johnson (1982) suggested that "in most cases, it does not appear feasible to establish an autonomous urban forestry agency, although there are some successful operations so structured." He found that of 12 cities studied, those with urban forestry functions within parks and recreation departments appeared to have more compatible goals, while those cities having forestry agencies in public works departments have more resources. Those cities whose urban forestry programs more strongly emphasize street tree care probably benefit from attachment to public works departments. Generalizations are difficult, however, as specific situations can make strong arguments for both cases.

Ordinances

Ordinances are legislative provisions adopted by local governments (counties, townships, or cities). "Ordinances manifest the concern of citizens and

211

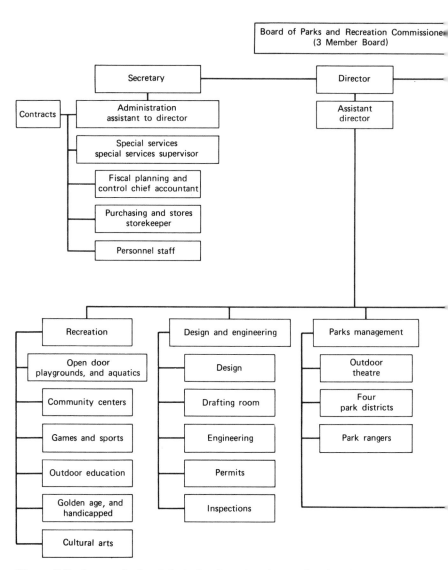

Figure 8.2 An organizational chart of a city park and recreation department having a policymaking board.

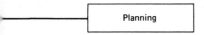

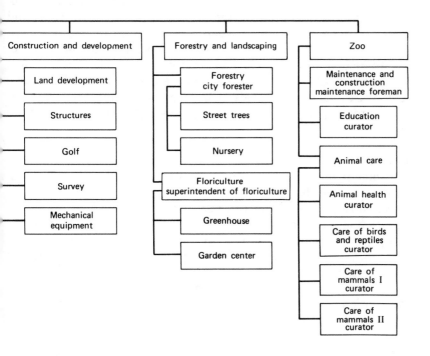

local officials who wish to capitalize on the beauty and comfort that urban trees provide" (Barker, 1978). While difficult to cleanly separate, ordinances are generally of two types: (1) those that pertain to trees on public property, primarily street and road sides, and tend to regulate planting and care; and (2) those that focus on trees on private lands within and peripheral to urban areas and reflect the increasing advocacy of special interests resulting from people's growing concern with the environment. We will deal at more length with the former, as they are more applicable to municipal forestry administration. The latter, however, the so-called "second generation" ordinances, deserve further comment. They often concern the preservation of trees on private lands—trees of historic significance, or trees preserved to maintain the wooded character of the landscape, often for property value and other economic reasons. In this category are ordinances requiring land developers to retain certain lands, often a percentage of the total land area being developed, as parks or open spaces. This category of ordinances is also related to land use zoning and other legal tools of urban planning.

City tree ordinances form the legal base for municipal forestry programs. They authorize and require city governments to care for trees and other vegetation in specific ways in order to provide for the health, safety, and well-being of residents. City tree ordinances vary from simple statements of authority and responsibility to lengthy prescriptions for management. For the most part, ordinances deal with streetside and other public trees. However, consideration is given to trees on private property that represent hazards to public well-being. They may include some or all of the following:

1. Definitions.
2. Establishment of a tree board or commission (composition, terms of office, duties, procedures).
3. Title to and responsibility for trees on public property.
4. Designation of a city forester and other officials (duties, authority).
5. Planting requirements (permits, official species, spacing, and location).
6. Maintenance (specifications and clarifying the responsibility of adjacent property owners).
7. Removal requirements and specifications.
8. Condemnation of trees on private property.
9. Requirements of public utilities and private tree care firms (licensing and insurance).
10. Prohibition of interference with commission or its agents.
11. Violations and penalties.
12. Specifications and standards of practice (may be written into the ordinance or attached as an appendix).

214

Detailed ordinances that require the intensive management of streetside and other public trees by adjacent property owners are often difficult to enforce. Knowledge of this difficulty has led many people to conclude that city tree ordinances are of generally little value. There is indeed little relationship between lengthy and detailed ordinances and the quality of municipal forest management. There is, however, a strong relationship between the quality of management and the quality of program administration. There should be no doubt that ordinances are necessary as well as basic to municipal forestry administration. They legalize commissions or boards and municipal forestry departments. They set forth requirements for public safety and authorize management. How well the forest is managed depends on how well what is authorized is provided for.

Ordinances must be specific to the needs of individual cities. Needs vary by city size, location, planting history, vegetation types and organizational structure, and by the requirements of state and federal governments. Needs change from one location to another because of geography and development. As a result, model ordinances, while beneficial as guides, are rarely applicable in their entirety (Unsoeld, 1980). The International Society of Arboriculture has prepared "A Standard Municipal Tree Ordinance" as a model that has been widely used (ISA, 1972). Appendix 2 contains two model ordinances—one used in the Kansas Community Forestry Program, and another developed for the Atlanta, Georgia, area. In both cases, the ISA ordinance was used as a base.

Financing

Municipal forestry programs cannot be effective without adequate financing. There are many services competing for public funds. Forestry programs have often a lower priority and thus are sometimes not sufficiently funded to carry out necessary management activities. It is important that good public relations be maintained and programs actively "sold" as to their benefits. Principal funding sources for municipal forestry programs include: (1) general funds; (2) capital funds; (3) municipal front-foot assessment; (4) special block or street assessments; and (5) permit fees.

Most municipal forestry programs are financed through general funds on the basis of a budget request made by the tree commission or the department under which the program is operated. Budgets are based on program needs and requests have to be justified. Major items are wages and salaries, supplies and new equipment, plant materials, operation and repairs, and equipment. These requests are submitted to the municipal governing body where final appropriations are made.

Capital funds are outlays for improvements that require the passing and issuance of revenue bonds. They most often involve specific needs such as tree planting after widespread disasters such as Dutch elm disease, tornados, or windstorms. In these cases, a special capital outlay is needed over normal operations. This need is presented to the populace and they decide by vote whether or not to issue the revenue bonds.

Another method of funding street tree programs is by property assessments. Here real property is assessed a certain amount for each foot of frontage. Resulting funds are then earmarked for personnel and equipment to perform the necessary management operations. Usually the right to levee such an assessment is based on state statutes. This type of funding does not have to be approved by public vote but by the action of the city council. It does, however, have to be voted on by the council from year to year.

Block or street assessment funding starts when a majority of the homeowners vote to subscribe to a tree management program. Once this majority has made this decision, the assessment is levied for all property owners on the block or street involved.

A permit method is often used in tree planting programs. Here the property owner pays for a tree to be planted by the city forestry department. It should be noted that the latter three funding sources (front-foot assessment, block assessment, and permit fee) often do not generate sufficient funds for good tree programs. When coupled with a general fund appropriation, however, they can provide for excellent programs.

In recent years, federal revenue sharing and other federal programs have provided additional sources of funds for municipal forestry programs. Table 8.1 gives sources of tree care funds as reported by Kielbaso and Ottman.

In smaller municipalities there may not be a sufficient public fund to conduct all operations. In developing programs for such communities, tree boards or commissions must look at three areas: (1) What can individual property owners do? (2) What can be accomplished by community action projects? And (3) what must require a city budget?

In most small towns, property owners are responsible for the planting and maintenance of street trees. It should thus be made as easy as possible for them to do whatever is necessary for street trees. This can be done by helping them select proper species, by providing accurate information, and by sponsoring projects where they are given opportunities to participate in tree management.

Nearly every community has organizations that want to become involved in city improvement programs each year—civic groups, service clubs, garden clubs, senior citizen groups, youth organizations, and the like. Forestry projects are often welcomed by them. Their efforts can accomplish

216

much if they are coordinated to assure that program goals and priorities are considered.

Individual or group efforts will not accomplish everything as some tasks require the commitment of public funds. Most community governments expect to spend money on a forestry program provided that they are assured that the money will be spent effectively. Thus, as mentioned earlier, the program or plan should define priorities. In most cases, top priority for public funds will have to be for the maintenance and removal of trees on public property—projects that do not lend themselves well to individual or community group efforts.

Urban Planning and Urban Forestry

There are three aspects to planning as related to the urban forest: (1) comprehensive urban planning; (2) land use planning; and (3) site planning. All are quite apart from management planning (an important aspect of urban forest management), and each exists in the political and policy environment of municipal and other local government administration. Comprehensive urban planning considers all the social, economic, legal, political and physical factors of the urban environment and represents an attempt to positively influence the future of urban areas. Yet, there are weaknesses in the planning process and problems inherent in dealing with planners (Thomson, 1983):

> *(1) Planning is poorly understood—it is a continuous, ongoing evaluative process which does not stop when the plan is adopted; (2) planning is an incomplete process (it does not include the resource base nor interaction between the resource and the people); and (3) planning is also a political process. It attempts to resolve conflicts between people, places and issues, and compromise is sometimes overlooked. In the final analysis, the citizen, not the forester nor the planner, must advocate urban forestry to the elected official who approves the plan.*

Land use planning is both an element in and product of comprehensive planning. And it is in the area of land use planning that urban forests and urban forestry are involved.

Land use planning is predicated on urban area change (growth, generally, but not exclusively), and it must be goal-oriented. The critical question is whose goals? The professional planner, the forester, and others will

217

TABLE 8.1
Sources of Tree Care Funds

Classification	No. of Cities Reporting (A)	General Funds No.	% of (A)	Special Frontage Tax No.	% of (A)	Tree Mill Levy No.	% of (A)	Road-Use Tax No.	% of (A)	Endowment No.	% of (A)	Vehicle Tax[a] No.	% of (A)	CETA[b] No.	% of (A)	Revenue Sharing No.	% of (A)	Gas Tax No.	% of (A)	DCBG[c] No.	% of (A)	General Forestry Grants No.	% of (A)	Other[d] No.	% of (A)
Total, all cities	1,040	982	94	8	1	9	1	23	2	18	2	17	2	90	9	88	8	49	5	9	1	34	3	98	9
Population group																									
Over 1,000,000	5	5	100	0	0	0	0	0	0	0	0	0	0	2	40	1	20	1	20	2	40	0	0	0	0
500,000–1,000,000	11	10	91	0	0	0	0	1	9	0	0	0	0	4	36	2	18	1	9	3	27	1	9	2	18
250,000–499,999	23	23	100	2	9	0	0	0	0	1	4	0	0	3	13	2	9	1	4	4	17	2	9	5	22
100,000–249,999	72	69	96	2	3	0	0	1	1	3	4	0	0	16	22	10	14	4	6	16	22	2	3	10	14
50,000–99,999	134	131	98	0	0	2	2	0	0	0	0	1	1	16	12	12	9	6	4	14	10	3	2	11	8
25,000–49,999	247	229	93	2	1	2	1	6	2	2	1	11	4	15	6	16	6	18	7	18	7	12	5	22	9
10,000–24,999	475	446	94	2	0	5	1	11	2	10	2	4	1	22	5	37	8	17	4	28	6	13	3	41	9
5,000–9,999	34	33	97	0	0	0	0	2	6	1	3	0	0	6	18	4	12	1	3	5	15	0	0	3	9
2,500–4,999	39	36	92	0	0	0	0	2	5	1	3	1	3	6	15	4	10	0	0	1	3	1	3	4	10

Geographic region																									
Northeast	251	245	98	1	0	2	1	5	2	2	1	0	0	7	3	6	2	1	0	8	14	6	7	3	18	7
North Central	330	300	91	6	2	7	2	13	4	9	3	11	3	38	12	39	12	25	8	44	13	22	7	39	12
South	221	214	97	0	0	0	0	2	1	5	2	3	1	24	11	20	9	19	2	15	7	1	0	17	8
West	238	223	94	1	0	0	0	3	1	2	1	3	1	21	5	23	10	19	8	18	8	4	2	24	10
Metro status																									
Central	226	220	97	3	1	2	1	4	2	5	2	2	1	38	17	24	11	11	5	40	18	13	6	26	12
Suburban	559	523	94	5	1	6	1	15	3	5	1	1	14	31	6	36	6	35	6	33	6	11	2	48	9
Independent	255	239	94	0	0	1	0	4	2	8	3	3	1	21	8	28	11	3	1	18	7	10	4	24	9
Form of government																									
Mayor-council	383	357	93	3	1	4	1	7	2	8	2	2	7	35	9	32	8	17	4	34	9	16	4	31	8
Council-manager	572	542	95	4	1	5	1	12	2	10	2	2	10	53	9	53	9	32	6	52	9	16	3	56	10
Commission	30	29	97	0	0	0	0	2	7	0	0	0	0	2	7	3	10	0	0	5	17	2	7	9	30
Town meeting	41	40	98	1	2	0	0	2	5	0	0	0	0	0	0	0	0	0	0	0	0	0	0	1	2
Rep. town meeting	14	14	100	0	0	0	0	0	0	0	0	0	0	0	0	0	0	0	0	0	0	0	0	1	7

Note: Percentages, when totaled, exceed 100 percent because some respondents indicated more than one source.

[a]Vehicle tax includes both state and local taxes on motor vehicles.

[b]Comprehensive Employment and Training Act.

[c]Community development block grant.

[d]Some of the other sources of tree care funds reported by cities were: tree sales, gas and weight taxes, state and federal categorical grants, and municipal water and utility charges. Cities also reported that some tree care funds were obtained from subdivision developer fees, city sales taxes, bond issues, and miscellaneous other sources.

all agree that urban residents should enjoy a high-quality living environment. Agreement at this level generally masks disagreement over a more detailed definition of what constitutes such an environment (Strong, 1978). The disagreement that must be resolved prior to setting realistic goals is likely to be over the components of what constitutes a high-quality living environment, the minimum value of each component, and the relative values of each. In such situations, people "go" where their interests are, and foresters not being exceptions are apt to place higher values on trees than others. It is perhaps not unreasonable to suggest that economic development is the strongest driving force behind planning, and while the esthetic values of forests and trees are held to varying degrees by all, they will be stopped at the unscalable wall of economic values.

Forestry's involvement with land use planning is generally extensive and concerns such inputs to the decision-making process as technical recommendations concerning site and vegetation relationships, potential site impacts, and the potential positive or negative impacts on the forest resource and forest industry. Involvement at this stage might also include identification and classification of individual trees or stands of trees of historic, economic, or cultural significance. As a matter of accuracy, it must be stated that the above is more idealistic than realistic, as forestry's input is not generally sought; simply because its role is not recognized or appreciated. It is our judgment that it is in this area foresters have an opportunity, and perhaps even an obligation, to influence the betterment of urban areas—a proactive opportunity little recognized by urban leaders, or by foresters themselves.

Forestry's most prevalent involvement in planning is in site planning and concerns topographical, soil, and vegetation evaluation; impact assessments; harvesting and marketing of products removed; protection during construction; planting plans, and management of residual forested areas. The significance and validity of forestry's role in this area are indicated by the conversion in Maryland alone of 325,000 acres, or 13 percent of the total of its forest land, to "urban" uses during the last decade. Moll (1982) suggests that it is critical to recognize urban areas with their myriad land uses as parts of the natural system:

> *Urban development is a land use and urban areas are part of the natural system. The function of the natural system will continue regardless of land modification. Developed lands become part of the natural system, no matter how incompatible it might be. We will always have weather and water will always flow downhill regardless of politics or social change. These existing natural systems are losing battles or at best, very expensive ones.*

220

Urban foresters are contributing information on natural sys-
tems during the development process. This application of forest
management principles to land planning is a concept long overdue.
Work with the developer–builder has been the most expedient
method of retaining forest vegetation. The developer–builder is the
action man in the development process. He takes a dim view of
the bureaucracy but will consider a good business offer. Show the
developer that retaining vegetation will increase his sales volume
and increase the selling price of his homes, and you will have his
ear.

Development is driven by economics, and forest protection, as suggested
above, is also motivated by economics. There are three economic "tragedies"
in failing to adequately consider and protect trees in development and construction: the increased cost of the property because of the presence of trees;
the cost of removal of trees inadequately protected during construction; and
the cost of replacement. It is to the avoidance of these tragedies that urban
forestry efforts are directed. Moll (1981) describes the process:

The techniques used in the process start when the development
concept is first conceived. During the first stage of the process three
things should happen at about the same time. The property should
be surveyed for boundaries and topography, the forest vegetation
reviewed and soil types evaluated. Much of this environment in-
formation can be determined by the forester as he walks the prop-
erty for the first time. The Soil Conservation Service can supply
information on soils from the soil series maps. Using the devel-
oper's concept plan, these professionals make some general state-
ments about the capabilities of the property.
During the second phase, the developer pinpoints some spe-
cifics about the type and quantity of the units to be constructed.
The engineer then discusses the design options with the forester.
The problems and potentials of the various types of vegetation are
weighed as part of the determining factor in the design. The en-
gineer prepares a plan with roadways, dwelling units and utilities.
The third stage takes the forester back to the field to review
critical parts of the plan; i.e., road crossings, streams, and steep
areas that are near valuable vegetation. If possible, the road center
line has been staked by the survey crew at this point. Recommen-
dations are made for minor location adjustments of roads, dwell-
ing units and utilities.
Finally, after the engineer has reviewed this most recent data,

the final plan is drawn. The forester returns to the field and marks the limits of clearing that will be required. He must determine at that point what trees will survive the construction and what measures can be taken to improve their survivability. Some species of trees may not be compatible with the proposed construction and should be removed. The vigor of others may be improved by thinning out the number of stems per acre.

A forester is brought in to scale the timber that must be removed to accommodate the construction operation. The value of the trees for lumber and other forest products is determined. The price that will be paid for a piece of timber is largely controlled by the skills of the forester at marketing timber. The profits gained from the timber sale are returned to the project budget and used to improve the landscaping and for revegetating the site.

Perhaps the best part of this coordinated effort is the cost to the developer, which is usually none. A considerable amount of money often remains for tree planting and revegetation. The resulting treed landscape will increase the sales value of the constructed units.

Site planning and follow-up involvement by urban foresters provide an excellent opportunity to prescribe the continuing management of not only landscape trees, but remaining natural stands of timber as well. Intensive multiple-use forestry is more applicable in these situations than any other areas of the nation.

BIBLIOGRAPHY

Barker, P.A., in "Some Key Aspects of Tree Ordinances," *Proceedings of the National Urban Forestry Conference,* Volume II, ESF Publication 80–003, SUNY, Syracuse, N.Y., pp. 724–729, 1978.

Bartenstein, F., "The Future of Urban Forestry," *Breaking New Ground in Urban Forestry,* The Pinchot Institute for Conservation Studies, USDA Forest Service, pp. 8–13, 1981.

Johnson, C., "Political and Administrative Factors in Urban Forestry Programs," *Journal of Arboriculture,* 8:160–163, June 1982.

Moll, G.A., "Forestry in Planning and Development," *Trends in Urban Forestry,* Vol. 18, No. 4, USDI National Park Service, Washington, D.C., p. 32, 1981.

Moll, G.A., *Land Development in Wooded Areas,* USDA Forest Service and Maryland Forest Service, Washington, D.C., p. 6, 1981.

Neely, D., and E.B. Himelick, eds., *A Standard Municipal Tree Ordinance,* International Shade Tree Conference, Inc., 2nd ed., 14 p., 1972.

Strong, A.L., "Planning and Management of the Urban Biophysical Environment," in *Proceedings of the National Urban Forestry Conference,* Volume I, ESF Publication 80–003, SUNY, Syracuse, N.Y., p. 46, 1978.

Thomson, M.A., "Making Urban Forestry a Part of the County and City Planning Process," in *Proceedings of the Second National Urban Forestry Conference,* American Forestry Association, Washington, D.C., 1983.

Unsoeld, U., *Analysis of Some Municipal Tree and Landscape Ordinances,* USDA Forest Service, Southeastern Area, 1980.

9
INFORMATION, EDUCATION, AND TRAINING

Information, education, and training programs are necessary for effective management of the urban forest. Property owners have the obvious need to be informed and educated concerning care of their trees. Private tree care firms benefit from an informed clientele and have special training needs for their personnel. In this chapter, we will look at information, education, and training largely from the standpoint of municipal forestry departments, as these departments have the largest responsibility and opportunity to influence management of the urban forest.

Information and Education

Information and education cannot be clearly separated. It can easily be argued that one is impossible without the other. The distinction was perhaps best made by the sage old school superintendent who said, "information is telling them what and education is them understanding why."

Information and education are necessary for three basic reasons: (1) to generate action by private property owners; (2) to gain acceptance of forestry programs; and (3) to create a favorable image of the department.

Generation of action by property owners

Municipal forestry departments have a responsibility to manage the total urban forest environment. As discussed in Chapter 6, this responsibility has limitations, as forestry departments have direct control over a limited part of the urban forest. They have direct control as prescribed by ordinance, policy, and tradition over city-owned areas such as parks and treelawns. However, areas of the urban forest in private ownership must largely be managed indirectly by the city. There is a limited opportunity for direct management as provided by ordinances that empower the city to remove hazardous trees and control epidemic insects and diseases, for example. However, most of what is done is by private owners. It is the objective of responsible city forestry departments to influence the forestry efforts of private owners in the best interests of the total forest environment. Therefore a vigorous information and education program is necessary.

The two objectives in such a program are: (1) to effect better general tree management by private owners; and (2) to make private owners more receptive to specific management needs. The former is self-explanatory, as proper selection, planting, pruning, and other maintenance are in the best interests of all the forest. In the latter, a certain practice might be desired by the city forestry department, but authority to require it is lacking. Examples are nonepidemic insect outbreaks where control measures would be desirable but not vitally necessary. In such cases, specific procedures would

226

be recommended to property owners. A specific example is on the Great Plains where cities often have a preponderance of Siberian elms, and infestations of elm leaf beetles are common. These insects rarely cause trees to die but make them unsightly and result in annoyance to residents. Many cities instigate spray programs on public trees and request that property owners spray their trees. Cooperation is often good and felt to be proportional to information and education efforts.

Acceptance of forestry program

Acceptance of the urban forestry program by the public is absolutely necessary, both from the standpoint of existence of the department and the implementation of specific practices. Programs of city government are largely proportional to citizen interest and demand. It is a political and practical reality that interest in a forestry program must be fostered, particularly in an environment of intense competition for public funds.

Urban forestry practices cannot be carried out without some degree of inconvenience to the public. Traffic must be rerouted, parked automobiles must be moved, and noise and dust are often raised. Obviously, if citizens are to tolerate these inconveniences, they must believe that the work is worthwhile. Public acceptance comes from an understanding of what is to be done and why it is necessary plus tangible, positive results of the work. This is closely related to another reason for information and education programs: a favorable image.

Creation of a favorable image

A favorable public image makes urban forest management easier. There is more acceptance of management practices if the department is respected. A favorable image is vital also to adequate budgets. This is a practical and political reality as suggested previously.

A favorable image results from doing the job well and having the public recognize that it was done well. Thus, the public relations program of a forestry department must be its total activity. It cannot be a veneer campaign to tell everyone what good guys they are. As suggested by Cook (1974), every organization has, by its existence, PR—good, bad, or indifferent—and any management decision transferred into action says something about the organization. A PR person or campaign does not "create" the department's image, but can articulate the policies and draw attention to the activities that qualify it for its reputation.

Fazio and Gilbert (1981) state that there are two distinct ways in which public relations in natural resource management is practiced. The first and most common is the constant day-by-day contact of all employees with the public, including such ordinary things as answering the telephone, greeting

227

office visitors, driving the organization's vehicles in traffic, answering questions, and generally conducting one's self in view of others. The second way is staff function, or a structured approach, that uses the services of one or more individuals who specialize in the practice. The latter could range from a large PR department, to specialized contracted services, to a part-time responsibility of a specific employee.

Public relations must permeate every function of a forestry department. Any specific activity (e.g., tree removal, pruning, or insect control) should be preceded by public announcements of what is to be done, why it is necessary, what it will involve, and how long it will take. The key effort should be in explaining the "what" and "why," based on the premise that understanding leads to acceptance. The limitations should be recognized, however, as there will inevitably be people who will not hear, will not understand, and may not accept.

Information and education methods and opportunities

Few municipal forestry departments are large and complex enough to have separate information and education divisions. Thus, PR must be the job of the total department. It takes two forms: (1) PR as a part of the everyday job; and (2) special PR efforts.

As suggested, every organization has, by its existence, PR. In the case of municipal forestry departments, equipment and personnel are noticed by everyone and the results of planting, pruning, and other work are also often visible. It is important that this visibility be in the best light. The ingredients of good on-the-job PR are simple: (1) good equipment—well maintained and clean; (2) good personnel—well trained and courteous; and (3) high quality work. There can be little doubt that courteous personnel in clean work clothes or department uniforms make a better impression. Likewise, clean, well-maintained equipment with the city seal and department name clearly displayed contribute to a good impression. The overall impression of a job well done, however, must result from just that—the job being well done.

The successful completion of a job and the correct impression that it was done well requires preliminary work. The job should result in as little inconvenience to the public as possible. Thus, it should be scheduled around events or activities that concentrate people and vehicles in a particular area. Through the media the public should be given notice about what is to be done, why, and how long it will take. If traffic is to be rerouted, alternate routes should be suggested. Specific notices to residents of the neighborhood where the work is to be done may also be necessary or desirable. It is obviously poor PR to surprise a neighborhood by showing up one morning, blocking off the street and starting chainsaws, trucks, and chippers.

Public relations communications must be two-way. The public must have access to the forestry department, and response to questions and problems should be quick. The public must be able to contact the department. The telephone number should be easy to find. There is no surer way of frustrating the public than to be unreachable by telephone, or to have inquiries responded to by unqualified personnel.

In addition to utilization of local news media for announcements, special projects are also desirable. Feature articles concerning forestry programs can be arranged. There is perhaps no better method of putting the good works of the department before the public than through a well-written and illustrated feature story. Most major newspapers have lawn and garden editors responsible for daily or weekly columns. These columns can be very helpful in recommending species and practices. A good working relationship with the media is extremely important. This relationship should be cultivated and maintained and should transcend personnel changes.

There are many clubs and organizations that can be helpful in the total forestry program. Civic and service clubs often seek environmental projects. Such projects can be designated by the city forester and their existence made known to club members. Projects should, of course, be a part of the city forestry plan and consistent with the overall program.

Garden clubs can be excellent ambassadors of city forestry programs. Their meetings offer an opportunity to explain the programs. Their projects in parks and other public areas can also make a significant contribution.

Many city forestry departments conduct special projects with schools, Scouts, and other youth groups. These can be informational or active projects such as tree planting or cleanup. Their long range purpose is to foster an appreciation of the urban forest.

Arbor Day ceremonies offer excellent opportunities for focusing attention on the urban forest. Ceremonies can be as simple or elaborate as desired, and can involve young and old people. Though normally associated with tree planting, Arbor Day can also stress the management of the total urban forest environment. Mayors, other city officials, and community leaders are generally happy to participate in Arbor Day ceremonies, often from a genuine concern for trees and often because of a "photo opportunity."

PR opportunities are as varied as the urban forester's imagination. Historic or heritage tree identification, specimen tree designation, and participation in the Tree City USA program are examples.* Each must be kept in perspective, however, and careful assessment made of their potential contribution to the urban forestry mission.

Communication with consulting urban foresters, commercial arborists, landscape architects, nursery people, university extension personnel, and

*For a description of the Tree City USA program, see Chapter 10, pp. 249.

other professionals is extremely important. In many cities, periodic meetings are held with such professionals to discuss common problems, update knowledge, and coordinate activities. All of them have an interest in managing the urban forest, and each can have a positive (or negative) impact on the city forestry program.

Commercial arborists have a large influence on the urban forest, as they are physically responsible for much of what is done. Arborists vary from highly trained competent professionals to unskilled laborers with little knowledge of the resource with which they work. A good rapport with commercial arborists is important in that it can provide the base for better management of the forest. Policies, standards, and specifications can be made available for licensing and contract bid purposes. Many cities require commercial arboriculture firms to be licensed for work within the city. In some situations, work on private property can be coordinated with management activities on public areas. This is particularly applicable in cases of insect or disease outbreaks.

Information and education (public relations) are key facilitators of municipal forestry programs. They are necessary to both short-range activities and long-range goals. They involve a communications flow from the forestry department to the public, and from the public to the department.

Training

The distinction between training and formal education is not clear. It can be argued that one is not possible without the other, or that they are one and the same. However, for the purposes of this section, we will attempt to separate them and will consider training as continuing education, beyond the prerequisite educational requirements of the job.

The training of urban forestry personnel is necessary for the following reasons:

1. To meet basic job standards.
2. To meet requirements of federal, state, and local laws.
3. To apply new technology.
4. To meet standards for certification or licensing.

Training to meet basic job standards

The necessity for training to meet basic job standards becomes apparent when the tasks and decisions of various urban forestry field operations are considered. Before considering these, it should be noted that a complex process of decisions and tasks is also necessary at the administrative level prior to field operations. Obviously, decisions must be made about priorities, locations, designs, information and education, and a host of other factors.

In the example of field operations, shown in Table 9.1, tasks are identified and decisions necessary to each are noted. The position level (C = crew, F = foreman, P = professional) at which each task is carried out and each decision made is indicated. Such a listing of task and decision levels helps to identify the expertise necessary to each. Training needs by levels can be determined then more readily. This listing also suggests the necessity of communicating decisions to the position level where the task is carried out. For example, the decision on the methods and materials to be used in staking and guying newly planted trees is made at the professional level. These decisions are communicated to the crew level where they become policy or standard operating procedure. The communication process is at the heart of personnel training.

Training can be formal or informal and can vary from structured courses to show and tell sessions between supervisors and employees. In-service training can be on a recurring schedule or during inclement weather. The advantage of scheduled training is that instructors and employees know when to expect it and can be better prepared. Training during inclement weather has the obvious advantage of efficient use of otherwise unproductive time. Most urban forestry organizations utilize both approaches according to need.

After training needs are clearly identified, there is the problem of finding qualified instructors. Baumgardt suggests to look first within the organization (Baumgardt, 1974). There might well be staff personnel qualified by work experience who could do a good job of training with proper reference materials and tools. Instructors can also be found through public agencies or professional or trade associations. Qualified consultants can also be engaged.

The content of the training course is extremely important. It must be relevant and presented at a level consistent with employee ability to comprehend. Most training in urban forestry work involves classroom explanation and outdoor demonstrations. Training of tree and shrub pruning perhaps illustrates this best.

Proper pruning involves not only a knowledge of tools and techniques, but an understanding of plant materials, plant growth, and response to pruning. Teaching this important job should include lectures, demonstrations, chalk-talk, the use of visual aids, and supervised practice. Employees should understand the "why" as well as the "how" of pruning. There are practical limits here, though, depending on employee ability and job needs.

TABLE 9.1
Tasks and Decisions in
Urban Forestry Field Operations

Task	Position Level of Task	Decisions	Decision Level
Tree Planting			
Designate spots where trees are to be planted	F P	Planting spots by species based on plan or site factors	F P
Supply trees to planting site	C F	Vehicle to be used, handling procedures, and on-site storage system	F P
Dig planting hole	C	Size of hole needed and digging equipment needed	C F
Prune roots if bareroot	C	Which roots to prune and how	C
Cut twine if balled & burlapped backfill	C	Planting depth and backfill material to be used	C F
Construct watering basin around tree	C	Width, depth, and shape of basin	C
Prune branches	C	Which branches to prune and how	C
Stake and guy tree	C	Method and materials	P
Wrap trunk	C	Method and materials	P
Mulch tree	C	Materials, method, and depth	P
Pruning and Repair			
Situation appraisal	C F	Species, treatments required, and equipment required	C F
Climb tree	C	Method and equipment	C F
Cut branches	C	Which limbs to cut, methods, and equipment	C F
Lower branches	C	Methods and equipment	C F
Paint pruning wounds	C	Method and materials	C F
Repair wounds	C	Which wounds to repair, method, and equipment	C F

Task	Description		
Brace slits or weak crotches	Methods, materials, and equipment	C	C F
Install cables	Locations, methods, materials, and equipment	C	C F
Dispose of debris	Method, equipment, and disposal site	C	C F
Tree Removal			
Designate trees to be removed	Dead, or otherwise in need of removal, how designate	F P	F P
Situation appraisal	Procedure—top and section, fell, etc.	F	F
Go aloft	Climb tree or use bucket truck (method and equipment)	C	F
Cut branches or portions of branches	Where to cut, method, and equipment	C	C
Lower branches	Methods and equipment	C	C F
Chip smaller branches	Material to be chipped	C	C
Load larger materials	Method and equipment	C	C F
Remove stump	Method and equipment	C	F
Dispose of debris	Disposal site	C	F
Insect and Disease Control			
Diagnose problem	Cause	P	P
Prescribe control	Control method, chemical, amount, timing, and equipment	P	P
Designate trees	Degree of potential damage and how to designate	F P	P
Mix chemicals	Amount of chemical and carrier	F	F P
Calibrate equipment	Amount of materials	F	F
Apply chemicals	Method, amount	C F	F
Clean equipment	Method, materials	C	F
Dispose of unused material	Method	C	P

C = crew.
F = foreman.
P = Professional.

Note. Tasks could be further divided. For example, cable installation requires hole drilling, holding with block and tackle, eye screw insertion, and cable attachment, each of which could be considered as tasks. A task is simply a part of an operation that is generally the responsibility of a single position level.

Training to meet requirements of laws

Training is necessary to meet requirements of federal, state, and local laws. At the federal level, two recent laws have had a profound influence on urban forestry: (1) the Federal Environmental Pesticide Control Act of 1972; and (2) the Occupational Safety and Health Act of 1971. Both laws are based on the government's right to protect the health and safety of the public.

The Federal Environmental Pesticide Control Act requires all state laws concerning pesticide use to conform to its provisions. Under law, applicators of pesticides must be certified. Certified applicators are either "private" or "commercial." Commercial applicators, including most arborists, are certified to apply "restricted use" pesticides. A pesticide is classified as "restricted use" if its dermal or inhalation toxicity presents a hazard, or if its use may cause unreasonable adverse environmental effects.

Standards for certification of applicators are prescribed, and provide that in order to be certified individuals must be competent in the use and handling of pesticides. Administered by appropriate state agencies, training and testing programs have been developed according to pesticide use category. The three categories that apply to most urban forestry operations are as follows:

Ornamental and Turf Pest Control: Includes commercial applicators that use or supervise the use of restricted use pesticides to control pests in the maintenance and production of ornamental trees, shrubs, flowers, and turf.

Right-of-Way Pest Control: Includes commercial applicators that use or supervise the use of restricted use pesticides in the maintenance of public roads, electric powerlines, pipelines, railway rights-of-way, or other similar areas.

Forest Pest Control: Includes commercial applicators that use or supervise the use of restricted use pesticides in forests, forest nurseries, and forest seed producing areas.

In most states, certification is given after the successful completion of written tests. Manuals and other materials for self-study are provided and training classes are scheduled. Provisions are also made for certification renewal, which require either class attendance or written testing.

The Occupational Safety and Health Act is an expression of the right of employees to a safe and healthful working environment. It requires em-

234

ployers to provide such an environment but protects their right to take disciplinary action against employees who violate safety and health rules.

Compliance with the Act requires that certain standards be maintained. In the area of urban forestry, the American National Standards Institute, Inc., has developed safety requirements for tree pruning, trimming, repairing, or removal (ANSI, 1972). Adherence to these requirements assures reasonable compliance. The International Shade Tree Conference (now the International Society of Arboriculture) was secretariat of the committee that developed the standards. They were developed in counsel with the Occupational Safety and Health Administration of the U.S. Department of Labor.

The six areas covered are: (1) general safety requirements; (2) electrical hazards; (3) mobile equipment safety requirements; (4) portable power hand tool safety requirements; (5) hand tool safety requirements; and (6) safe work procedures. These standards and requirements form a solid base for a safety program. Such a program begins with administrative policies, involves standards for types and condition of tools, equipment, and materials, and culminates with requirements for use.

Compliance with state and local laws also often requires training. Licensing of persons or firms who do urban forestry work may be required by state law. An example of such requirements will be discussed later in this chapter. Licensing is often required by municipal ordinances also and may be based on oral or written tests, or acceptance of state licensing or certification in states where such programs are in effect. Training for meeting licensing requirements usually involves studying a manual or book prior to a test. In a few cities, however, practical training courses are presented.

Training to apply new technology

The continuing development of equipment, materials, and procedures makes apparent the need for continued training. In recent years there has been rapid advancement in tree planting technology. New equipment and materials for digging, transporting, and transplanting trees have come on the scene. An array of new chemicals is now on the market with complex systems for their use. There are new materials and techniques for fertilizing, watering, aerating, controlling insects and diseases, and new equipment for application.

The necessity of training to apply new technology is well illustrated by the recent development of new Dutch elm disease control systems. This development also illustrates the training needs required by law. In early 1972, the Environmental Protection Agency gave label approval for the fungicide, benomyl, to be used in Dutch elm disease control. The label approved either a foliar spray or trunk injection. The spray was to be applied with a mist blower after trees reached full leaf in the spring. Trunk injections were also to be made after full leaf. Equipment systems were developed for injection. The EPA label restricted benomyl use to "trained arborists," but ironically did not define the term or say who was to train them. Training did occur, however; it was conducted largely by chemical company sales personnel and by representatives of manufacturers of injection systems.

Benomyl was successful when used under proper conditions. Its effectiveness was restricted by a frequent lack of understanding of limiting factors. Effectiveness was further restricted by benomyl's frustrating insolubility and frequent resistance to uptake in trees.

A soluable form of benomyl was developed and tested from 1973 to 1975. This material, Lignasan BLP, was given label approval in the spring of 1976. The major educational need of Lignasan BLP appears to be an understanding of the proper conditions for its use. This is a need of both applicators and consumers. There is also the continuing need to train users in the physical handling and use of the material.

Training to meet standards for certification or licensing

The certification or licensing of arborists and others concerned with the urban forest is often required by state or local laws. Although this discussion might properly be included in the previous section, it is considered separately because certification or licensing is generally professionally motivated. Certification can be required by law or can be a voluntary program that is sponsored and administered by an association or other organization. The requirements for certification or licensing usually result from a desire to be more "professional." This desire may be based on a combination of profit oriented and consumer protection reasons. It is also based on a genuine concern for the quality of work done in the urban forest.

Unsoeld (1980) reported that

> Licensing and registration often lead to several conclusions. Proper licensing raises arboricultural standards, breeds professionalism, and improves public attitudes. However, acquiring a license costs money, enforcement is not free, a fair and comprehensive test must be developed, the unavoidable paperwork requires additional personnel, and those unlicensed raise cries of competitive advantage and encroachment on individual freedoms. In some cases, politics or economics cloud the development of an adequate law, or place its administration in the wrong department. Anti-regulation attitudes often prevent passage. To cover administrative costs, fees are, occasionally, set too high, resulting in Court reversals. Arborist associations sometimes circumvent legal problems by creating their own voluntary certification programs. They set the fees

236

as high as they choose and make the test as tough as they please, but occasionally discover that much of the public is unaware of their efforts. To improve public awareness, lists of registered arborists are often distributed, but sometimes fail to reach those who need them most.

The Kansas Certified Arborist program is an example of a voluntary approach without legislative authority. The program is sponsored by the Kansas Arborists Association, an organization of professional and lay people concerned with shade and landscape trees. As required by the Association's bylaws, certified arborists must meet the following standards:

1. Membership in the Kansas Arborists Association.
2. Successful completion of a course in arboriculture as prescribed or approved by the KAA.
3. A minimum of two years experience as a practicing arborist.
4. Practice of a code of ethics.
5. Payment of an annual certification fee.
6. Possession of insurance in the amount of $50,000 property damage and $100,000 personal liability or ample amount to meet current needs.

The Association does not certify competency. To do so would be extremely difficult. It does, however, certify that the holder meets the above standards, which should reasonably help to provide competency. The program's strength is that it requires the "successful" completion of a prescribed or approved course in arboriculture and two years' experience as a practicing arborist.

An annual arborist training course is sponsored by the Kansas Arborists Association and Kansas State University. The week-long course includes both classroom and field work. Of necessity, it deals with fundamentals and is designed to serve as a basis for further development (Figure 9.1). In addition to the training course, continuing membership in the Association is required. This provides an opportunity for further training at conferences and field days.

The New Jersey "Certification of Tree Experts" program is an example of a program based on state legislation. A bureau of tree experts is provided for within the state's department of environmental protection. The bureau administers the program and certifies "tree experts" who meet the following qualifications:

1. U.S. citizen (or declaration of citizenship intent, and legal resident of New Jersey).
2. Over the age of 21 years.

237

Kansas Arborist Training Course Outline

Monday:
9:00	Introduction
	Welcome, course outline, requirements, crew assignments, information, questions
9:30	Definitions, history
9:45	The tree, its parts, and how it grows
11:00	Trees and the soil
1:00	Tree forms and sizes, selection and use
2:00	Field trip—soils, space and other site factors, tree identification methods
5:00	Problem and reading assignments, adjourn

Tuesday:
8:00	Quiz
9:00	Pruning (classroom)
	Why Prune?
	Pruning response
	Pruning methods and practices
	Pruning for structure
	Training young trees
	Pruning at planting time
	Pruning mature trees
	Pruning shrubs
1:00	Field work (by crews)
	Tree planting
	Pruning of newly planted trees
	Pruning of small trees
	Shrub pruning
5:00	Problems and reading assignments, adjourn

Wednesday:
8:00	Quiz and critique
9:00	Demonstrations and field work
	Pruning large trees
	Aerial equipment
	Ropes, saddles, knots
1:00	Safety
2:00	Field work (by crews)
5:00	Problem and reading assignment, adjourn

Thursday:
8:00	Quiz and critique
9:00	Cabling and bracing
	Bark tracing
	Wound repair
	Drain tubes
11:00	Movies
1:00	Field work (by crews)
	Cabling and bracing
	Bark tracing
	Wound repair
4:00	Adjourn
6:00	Woodpecker barbecue

Kansas Arborist Training Course Outline (continued)

Friday:

8:00	Tree fertilization
9:00	Tree insects
10:00	Tree diseases and physiological problems
11:00	Tree fertilization and spraying demonstrations
1:00	Tree problem diagnosis
3:00	The professional arborist
	Arborist code of ethics
	Proper tree care alternatives
4:00	Quiz, critique and wrapup

Figure 9.1 Kansas Arborist Training Course Outline.

3. Good moral character.
4. Four years of college education, preferably forestry, agriculture or equivalent, or sufficient professional experience in tree care to waive college education, or five years continuous experience immediately preceding the date at which an individual applies to be considered a tree expert.
5. Successful passing of examinations in the theory and practice of tree care, including such courses as botany, plant physiology, dendrology, entomology, plant pathology, and agronomy.

Examinations are held at least once a year and may be oral, written, or both. Applicants who fail to pass may be reexamined at the next scheduled examination period. The program does not restrict anyone from working in the field of tree care. It provides a means by which individuals may be examined by the bureau of tree experts, judged competent, and authorized to use the term, "certified."

BIBLIOGRAPHY

Baumgardt, John, "How to Train a Pruning Crew," Grounds Maintenance, Vol. 9, No. 2, February, 1974.

Cook, John, *John Cook's PR Without the BS,* John Cook and Company, Phoenix, 1974.

Fazio, J.R., and D.L. Gilbert, *Public Relations and Communications for Natural Resource Managers,* Kendall/Hunt Publishing Co., Dubuque, Iowa, pp. 7–8, 1981.

Unsoeld, U., *Certification/Licensing/Registration of Arborists,* USDA Forest Service, Southeastern Area, 1980.

239

10
URBAN
FORESTRY PROGRAMS,
SUPPORT ORGANIZATIONS,
AND RESEARCH

There is no single government agency, private corporation, association, or other organization directing or coordinating urban forestry in America. Interests and ownerships are far too diverse for a single umbrella. Rather, urban forestry involves a wide range of public and private programs, related organizations, and specific research efforts. In spite of a lack of central direction, or perhaps because of it, great progress is being made because of the willingness of the people involved in urban forestry to both share and seek knowledge.

For the purpose of this discussion, an urban forestry program is defined as the total effort of an organization, agency, or business toward urban forestry. Examples are city forestry programs, urban and community forestry programs of state forestry agencies, and urban forestry efforts of consulting forestry firms. City forestry programs are most familiar and have been discussed in varying detail in previous chapters. The role of state agency urban foresters and urban forestry consultants is less well known and will be explored further.

State Forestry Programs

An amendment to the Cooperative Forest Management Act in 1972 authorized the U.S. Forest Service to assist state forestry agencies in developing programs in several states. The state forestry agencies of Georgia and Missouri placed urban foresters in Atlanta, St. Louis and Kansas City in 1967 and 1968. They have developed programs responsive to the requests of urban residents and serve a clientele holding potential support for their agencies' total programs. Their work is largely educational; it includes work with mass media, schools, trade associations, garden clubs, civic and service clubs, and youth groups. Assistance is provided to property owners, tree service companies, land developers and local units of government including park, cemetery, and planning boards. Demonstrations and training schools are sponsored for city forestry department personnel and others responsible for the maintenance of the urban forest. Similar programs have since been developed in other large metropolitan areas of the nation.

The Kansas community forestry program

Kansas has developed a forestry program for small towns that has become a model for other states. The Kansas Community Forestry Program began in 1971 in response to numerous appeals from small towns hit hard with Dutch elm disease in the late 1960s. With technical assistance provided by the Department of Forestry at Kansas State University, the program provides for the creation of local tree boards or commissions that develop and administer forestry programs for their towns.

To be effective on a continuing basis, tree boards must be legally constituted with the authority to develop and implement programs (Appendix 2, page 265). Ideally, board members should be dynamic community leaders who are able to make and carry out decisions. A knowledge of trees is important, but perhaps secondary to leadership ability and interest.

Before programs can be developed, tree boards must know the public tree situation in their towns. With assistance from state forestry department personnel, an inventory of street and park trees is taken. Trees are inventoried by species, size, age, and condition class, and from these data species and condition class percentages are computed.* The inventory serves as a planning base from which planting, tree removal, and maintenance needs can be determined. Priorities are then determined and program goals set. It then becomes necessary to identify what is needed to reach the goals. For example, if a tree board gives priority to a street tree planting project, the following questions must be answered: (1) What species will be made available? (2) How will orders be taken? (3) How will the program be publicized? (4) How will payment be arranged? (5) How will planting spots be located? (6) How will trees be planted and maintained? And (7) what ordinances are desirable to regulate street plantings? Similar questions must be asked for tree removal and maintenance projects. The answers to these questions constitute annual or project work plans.

Ideally, tree boards should serve in a planning and advisory capacity, with qualified city employees to physically implement work plans. Realistically, however, most small towns do not have employees with either the responsibility or qualifications for such work. Hence, in most towns tree boards actually administer and implement the program. Board members purchase and distribute trees, arrange publicity, mark trees for removal, stake planting sites, and the like. This approach has the obvious weakness of being dependent on the board members' interest. However, in small towns with limited resources, it is perhaps the most practical approach.

Provision of technical service is vital to the Kansas community forestry program. At the outset, few towns have the capability or available funds for the development of landscape plans, dead and diseased tree marking, planting site location, or insect and disease diagnosis. If work plans are to be carried out, these services must be made available. Urban foresters and a landscape architect from the state forestry department provide such services. Private landscape firms also play a role.

A sample inventory summary is shown in Chapter 3, pages 39, 43, 44.

The staff landscape architect makes a strong contribution to the program. Many towns give high priority to the development of park and recreation area plans or the landscaping of other public areas such as city squares, highway entrances, and central business districts. The first determination is whether or not funds are available to contract with a private landscape firm. If not or if the project is too small to be economically feasible for a private firm, the service is provided by the state forestry department. If funds are available, a list of private contractors is furnished to town officials.

Many plans are developed on site by simply staking planting locations. This is a direct, positive way of getting plantings in the ground. It provides for efficient use of the urban forester or landscape architests' time and is applicable in cases where planting funds are available. In other cases where implementation funds must be budgeted, drafted plans are necessary.

In addition to assisting the tree board organization, program development and implementation, Kansas urban foresters provide training for tree board members, city employees, and private arborists. All are involved in the Kansas Arborists Association—tree board members, private arborists, and foresters.

As of this writing, there are 130 towns in Kansas that have developed comprehensive community forestry programs. They range in populations from 151 to 28,000—the average is 4,756. Some are extremely successful. Some flounder, and some have failed. Those that are successful have common ingredients; (1) a supportive and positive attitude of city officials and residents; (2) legal sanction and authority—a part of city government; (3) a readily available source of accurate technical information and assistance; (4) projects that are practical and consistent with community abilities; and (5) consideration of the rights and wishes of individual property owners. It is our conviction that these same ingredients are necessary in any city in the nation regardless of size.

Virtually every state forestry agency offers some form of urban forestry assistance. Programs vary greatly, depending on interest and budgets. Involvement may be comprehensive or limited to responding to requests by individuals or groups. Urban forestry programs take many forms and may involve any combination of the following:

Audience	Involvement
Local government (city) commission, council, etc.)	Counsel of program organization and development—for example, ordinances and sample costs
Tree boards, commissions (action or advisory)	Forestry program development, project planning, information, and training

Planning boards, commissions, and authorities	Tree protection and environmental assessments, for example
City forestry departments	Information, training, and project planning
Private developers	Tree protection, planting information, and environmental protection
Private arborists	Information, education, and training, (e.g., schools and workshops)
Clubs and organizations	Information and project assistance
Property owners	Information and education (media, direct contact, demonstrations, and workshops)

Urban forestry programs of state forestry agencies are well summarized in a 15-minute movie *The Urban Forest*, produced at Kansas State University in cooperation with the USDA Forest Service. The movie shows program examples in various states and appeals to city officials for action. Development and production of the film were part of an urban forestry program that obtained technical input from state forestry agencies, the Urban Forestry Committee of the National Association of State Foresters, and personnel of the USDA Forest Service.

Consulting Urban Foresters

An increasing number of private consulting firms and individuals offer urban forestry services. These services go beyond traditional arboriculture and cover a wide range of applications of conventional forestry tailored to meet urban situations. In 1974, there were 15 foresters in the United States offering urban forestry services. The current number is unknown. Our observations suggest, however, that this number has increased greatly since then. Evidence is found in journals and other publications carrying notices of consulting services. Also, major arboricultural firms have expanded to include urban forestry in their listing of offerings.

Services of consultants vary according to the localized demand and expertise available. Services are offered to municipalities, other governmental units, corporations, and individuals; they include the following:

• Information on tree care
• Street and park tree surveys and inventories
• Program development counsel
• Insect and disease inspections and diagnosis

245

- Tree care contract service including specifications and inspections
- Land development tree surveys
- Vegetative management plans
- Studies for environmental impact assessments and statements
- Advice on selecting, planting, and moving trees
- Evaluation reports on losses from fire, storms, accidents, and other causes
- Tree problem diagnosis
- Expert testimony in court cases

Support Organizations

Various organizations, by virtue of their dedication to trees and the environment, give support to urban forestry (Unsoeld, 1980). Notable among them are the International Society of Arboriculture, the National Arborist Association, the Urban Forestry Working Group of the Society of American Foresters, the National Arbor Day Foundation, the National Urban and Community Forestry Leaders Council, state urban forestry councils, state arborists associations, and a great number of volunteer organizations.

International Society of Arboriculture

Formerly the International Shade Tree Conference, the International Society of Arboriculture has the following purposes:*

> To promote and improve the practice of professional arboriculture.
>
> To stimulate greater public interest in the planting and preservation of shade and ornamental trees.
>
> To promote public education to develop a greater appreciation for trees of shade and ornamental value, and to promote cooperation in the conservation of trees and in the beautification of the countryside.
>
> To recommend and uphold a Code of Ethics established to maintain a high level of practice by those engaged in the profession.
>
> To initiate and support the scientific investigation of programs concerned with arboriculture and to publish the results of such investigations.
>
> To sponsor an annual meeting devoted to the exchange and presentation of information of interest and value to professional

---

*International Society of Arboriculture, *An Invitation to Join the International Society of Arboriculture*, 3 Lincoln Square, Urbana, Illinois.

arborists and others interested in improving tree planting and maintenance practices.

To afford the producers of materials, services, and equipment of value to arboriculture an opportunity to advertise, exhibit, and demonstrate their value to arborists.

A wide range of tree interests is represented by the nearly 3,000 members—commercial arborists, municipal arborists, urban foresters, research scientists, park and grounds superintendents, consultants, city tree commission and board members, landscape contractors, landscape architects, nursery people, educators, and others.

The ISA has 10 chapters organized on a regional basis. At the national level, much of the work is done in subject matter committees such as those concerned with safety, pesticides, shade tree evaluations, and urban forestry. Other committees involve administration and affiliated associations. These associations include the Municipal Arborists Association, Utility Arborists Association, and Arboricultural Research and Education Academy. Other related or affiliated organizations are the American Society of Consulting Arborists, various state arborists associations, and associations of shade tree commissions.

The ISA publishes a monthly *Journal of Arboriculture* containing articles by researchers and practitioners. Education materials such as slides and films are made available. In addition, a "Standard Municipal Tree Ordinance" has been published, and a shade tree evaluation formula developed.

Society of American Foresters

The Society of American Foresters is the primary organization of professional foresters in the nation. The objectives of the Society are to "advance the science, technology, education, and practice of professional forestry in America and to use the knowledge and skill of the profession to benefit society" (SAF, 1975). Professional foresters have long been involved with the urban forest environment. The major emphasis of Society members, however, has naturally been on matters relating to forest resources beyond urban environs. In 1972, working groups were formed within the Society. The Urban Forestry Working Group provides a focal point for foresters interested or involved in this area. Its objectives are (SAF, 1972):

1. Establish liaison with such associations as the International Society of Arboriculture and the Society of Municipal Arborists to:
   a. Exchange information on purposes and objectives of the respective organizations.
   b. Avoid duplication of effort in setting up training meetings and/or

247

conferences or other educational projects where interests and purposes are common to the respective organizations.

c. Exchange results of research and field testing.

2. Represent the interests and needs of urban-oriented foresters as a means of holding their interest in the society.

3. Serve as the advisory branch of the Society of American Foresters in the field of urban forestry.

4. Determine educational needs and career opportunities in urban forestry and encourage educational institutions to develop a curriculum in urban forestry or to improve existing courses.

5. Communicate with public and private planning agencies to encourage their employment of professional foresters.

6. Coordinate programs and plans with other appropriate Society working groups.

7. Serve as consultants to sections of the Society of American Foresters when assistance is needed to set up regional technical conferences.

8. Identify areas of needed research in urban forestry.

In addition to the setting of these objectives, these projects of immediate priority were identifed:

1. Prepare a directory (listing by states) by giving the names of professional foresters (public or private) working in the field of urban forestry [completed (SAF, 1974)].

2. Prepare a compendium of state laws and municipal codes on urban/community forestry.

3. Prepare a bibliography of major publications in the field [completed (Andresen, 1973)].

4. Collect and distribute job descriptions and salary schedules currently in use for urban forestry positions.

5. Prepare a policy statement for the Society on the importance of urban forestry including a definition of urban forestry in the statement (see Chapter 1).

National arbor day foundation

The Arbor Day Foundation is a nonprofit corporation directed by a board of trustees made up of conservation and business leaders in the nation. The purposes of the Foundation are to:*

> *Properly and officially promote the observance of Arbor Day each year.*

*National Arbor Day Foundation, *Your Opportunity to Improve Your Town*, Arbor Lodge 100, Nebraska City, Nebraska.

Create an awareness and appreciation among all peoples through all forms of communication of the fundamental role that trees play in man's day-to-day existence.

Endorse, support, or otherwise implement education programs that will stimulate and inspire youth to better understand the bounty and joy of trees.

Recognize achievement among all elements of society through an annual Awards Program for contributions made to the understanding, appreciation, conservation and wise use of trees.

Initiate programs that encourage the planting of trees and create an awareness of those resource programs that will assure the perpetuation and growing abundance of this basic resource.

Establish and maintain a National Arbor Day Center at Nebraska City, Nebraska, the birthplace of Arbor Day. The basic purpose of the Center will be to educate Americans on Arbor Day, the resource of trees, and tree planting . . . past, present and future.

A specific urban forestry activity of the Arbor Day Foundation is the "Tree City USA" program. Tree City USA is designed to recognize cities and towns that are effectively managing their urban forests. It is also designed to encourage the implementation of local forestry programs by meeting the following standards:

1. A legally constituted tree body (a department, board, commission, or other authority) with statutory responsibility for developing and administering a comprehensive city forestry program.
2. City ordinances or statutes providing for tree planting, maintenance, and removal according to proper urban forestry principles.
3. An active comprehensive forestry program, supported by a minimum one dollar per capita public funds.
4. A formal Arbor Day proclamation by the mayor and a commemorative tree planting each year.

The program was begun in 1975 and is administered locally by cooperative state forestry agencies. Its impact as an incentive for development of local municipal forestry programs has been significant. There are currently 364 cities and towns designated as "Tree City USA" in the nation.

National Urban and Community Forestry Leaders Council

The National Urban and Community Forestry Leaders Council is affiliated with The American Forestry Association and is the primary vehicle for AFA's emphasis in urban forestry. Founded in 1981, the Council is a coalition of individuals from communities, government, industry, and professional or-

249

ganizations with a dedicated interest in urban forestry. The Council strives to:

Foster understanding and to promote the concepts of urban and community forestry.

Provide a unified national support base for urban and community forestry.

Create a climate that fosters urban and community forestry among urban and natural resources organizations.

Communicate the benefits of urban and community forestry to local, state, and national leaders.

Recognize those making outstanding contributions to the field of urban and community forestry.

The Council operates with an informal structure, with most of its projects coordinated by a "core" group made up of national leaders in urban forestry from the USDA Forest Service and Extension Service, American Forestry Association, and forest industries. Among its notable achievements are publishing the "Forum," a detailed newsletter; creation, development, and promotion of "Spunky Squirrel," AFA's urban forestry symbol; development of an urban forestry awards program; and coordination with AFA of sponsorship of the Second National Urban Forestry Conference. DeBruin (1982) stated the premise on which the Council is built:

The Council and AFA believe that U&CF is the gateway to forestry of the future and to the development of a sound conservation ethic. It is further believed that urban forestry can provide the means to help solve some of the deep-rooted problems troubling our cities and communities. Urban forestry can:

Improve the quality of life in our cities.

Reduce noise.

Reduce air pollution.

Reduce climatic extremes.

Conserve energy.

Provide beauty and shade.

Increase property values.

Provide wood and fiber.

Provide first-hand opportunities, within living and working environments, for interaction between people, trees, and other natural resources.

Control water runoff and replenish aquifers.

Provide urban habitats for a variety of wildlife.

Improve water quality and fishing opportunities in local streams.

In short, sound management of forests and other natural resources in heavily populated areas will not only make our cities better places in which to work but also more desirable places in which to live.

Volunteer organizations

Virtually every large city in the nation has one or more volunteer organizations dedicated to tree planting, urban beautification, or other activities relating to urban forestry. These organizations tend to arise in response to a perceived need by individuals or groups. They come and go according to project needs and the interests of their leaders. They might be formed as a project or issue-oriented activity of a parent group, such as an Arbor Day committee of a service club, or might be new organizations created to serve an urban forestry need. Volunteer organizations are both physically and advocacy-oriented. They physically carry out projects (most often tree planting), and they also often speak and even lobby for trees. A complete list would not be possible here, but such organizations could be grouped as follows: neighborhood associations; parent group committees; specific support groups, such as "Friends of the Park"; and general urban forestry organizations.

A classic example of the latter is the "TreePeople" of Los Angeles. TreePeople is a nonprofit corporation made up primarily of volunteers and nonforestry professionals. It started in the late 1960s with the efforts of its current leader, Andy Lipkis (1983), to replant a portion of the smog damaged San Bernardino National Forest and has since grown into an organization of thousands of urban volunteers involved in tree planting, tree maintenance, and educational programs. TreePeople undertook a massive effort to involve the entire Los Angeles population in planting and maintaining one million new trees to ready the city for the 1984 Olympics.

Another volunteer program, also in California, but of a different nature, is TREES, an acronym for Trained Residents Extending Educational Services. Sponsored and directed by the Cooperative Extension Service of the University of California, this program trains individuals from selected communities in the principles of urban forestry. The trained volunteers then conduct urban forestry programs in their own neighborhoods (Bone, 1983). The TREES volunteers complete 27 hours of classroom and field training and agree to provide a minimum of 27 hours of direct service. Many have given literally hundreds of hours and been instrumental in forming other volunteer groups.

Volunteer organizations are vitally important elements of the urban forestry effort in individual cities and in the nation. Volunteer efforts, though, must be coordinated with, or have access to, professional programs. As with the TREES program, they can add great strength to professional organizations. A major obstacle, however, can be the professional organization itself. This was well stated by Appleyard (1978):

> *Programs that encourage community planting and care of trees could enable the public to learn more about trees, to respect and care for them, to come closer to them, and to enjoy their manifold meanings. The principle blocks to such involvement may be the professional tree management and park departments themselves. Too often parks and recreation and public works departments are oriented to professional control, minimal maintenance, and minimal public 'interference,' rather than sharing tree management with the public citizenry.*

Lipkis (1983) stated the importance of a close relationship between volunteers and government agencies:

> *TreePeople has learned how to work with agencies in a more effective way, taking time to build trusting relationships. It also seems very important that agencies prepare themselves to better deal with volunteer groups and the public. If agencies are more supportive of volunteer efforts and help supply guidance, training and leadership, they will find the community a very good resource to support them in their work, both politically and practically.*

Research

Research in urban forestry is a dynamic need, because the changing face of the urban environment presents challenges requiring new trees and new technology. Research relating to urban forestry is conducted primarily by industries, federal agencies, state and local government agencies, colleges and universities, private research corporations, institutes, and academies.

252

Research is funded largely by these organizations, but may also be supported or endowed by private contributions, trusts, foundations, or other sources.

Research related to urban forestry by federal agencies is conducted primarily by the USDA Agricultural Research Service and the Forest Service. Research by the Agricultural Research Service is generally in the area of shade and windbreak tree improvement. The Forest Service is involved in research specific to urban forestry and also conducts research in other areas that have application to urban forestry.

Authority for USDA Forest Service research is granted by the following legislation:

- McSweeney-McNary Forest Research Act of 1928.
- Agricultural Experiment Stations Acts of 1955.
- Cooperative Forestry Research Act of 1962.
- Whitten Act of 1956.
- Public Laws 85-934 and 89-106 (Research Grant Acts).

These laws authorize in-service research and also provide for cooperative programs with universities and other institutions. USDA Forest Service research specific to urban forestry was summarized by Moeller (1981):

Development methods to manage urban vegetative systems and integrate urban forest planning with comprehensive urban development planning by: identifying criteria and methods to establish, maintain, and protect urban forest vegetation; developing technical information systems needed for urban forest planning and management; and developing methods to assess landscape values and integrate these values into planning.

Develop strategies to advise decision makers on managing urban forest resources for recreation by: developing methods to assess recreation needs of urban residents; evaluating the capabilities of urban forest resources to meet these needs; and designing and implementing management programs to meet urban recreation needs.

Develop more effective methods to integrate scenic, recreation, and wildlife values in the planning and management of urban forests by: quantifying scenic values of urban forest landscapes; identifying amenity benefits from urban forests (including recreation); improving wildlife habitat for nonconsumptive uses; and developing techniques for integrating scenic, recreation, wildlife, and other amenity values of urban forests into the urban land planning process.

Develop ways to integrate planning and management of urban forests with urban planning by: determining quantitative influences of various types and intensities of urban development and ecological functioning on urban forests; and organizing urban forest management and planning systems that mesh with comprehensive urban development and planning.

Develop urban forestry management guidelines to improve the physical environment by: determining ways in which urban forests can be manipulated to influence micro- and meso-climate, noise, and water values; and developing knowledge on types of forest vegetation most suitable for metropolitan environments in the Northeast.

Develop methods to establish, manage, and protect urban forest vegetation and to assess urban forest benefits by: determining the nature and magnitude of social and economic benefits from urban vegetation; and developing criteria for selecting, establishing, maintaining, and protecting urban forest resources in the Southeast.

Moeller states also that:

> *future research will treat the urban forest as a system that is capable of producing multiple benefits. Specific areas of emphasis will include: (1) development of multiple use management strategies for urban forest vegetative systems; (2) assess emerging urban forest resource demands under conditions of energy scarcity; (3) design energy efficient urban forest management programs; and (4) expand research on biological/physical components of the urban forest system.*

Research by private industries is primarily product-oriented and involves the development and testing of plant materials, equipment, and supplies. Systems for the use of equipment and supplies are also researched. Selection, breeding, and testing of "new" trees by nursery people is a prominent example of private industry research. From this work has come an array of new plant varieties displaying special form, color, and other qualities in recent years. There is continuing research by chemical companies in insect and disease control, tree fertilization, and growth control. The recent development of injection capsules for insecticides, fungicides, fertilizers, and other chemicals is a result of research of application systems. Large equipment such as mechanical tree transplanters, aerial buckets, and sprayers are obvious products of research and development.

A great amount of research is done for private industry by university scientists working under grants and contracts. Such arrangements can be particularly important in situations requiring the testing of materials or the methods for application in different areas of the nation.

254

BIBLIOGRAPHY

Appleyard, D., "Urban Trees, Urban Forests: What Do They Mean," *Proceedings of the National Urban Forestry Conference,* Volume I, ESF Publication 80–003, SUNY, Syracuse, N.Y., 155 p., 1978.

Bone, P., "Master Gardener and TREES—The Volunteer in Urban Forestry," in *Proceedings of the Second National Urban Forestry Conference,* The American Forestry Association, 1983.

DeBruin, H.W., "Urban and Community Forestry: The Time Is Now," *American Forests,* The American Forestry Association, Washington, D.C., March, 1982.

Lipkis, A., "One Million Trees for the 1984 Olympics in Los Angeles," in *Proceedings of the Second National Urban Forestry Conference,* The American Forestry Association, 1983.

Moeller, G.H., "Forest Environment Research, Forest Service Urban Forestry Research Program Summary: Past, Present, and Future," USDA Forest Service, unpublished, 1981.

Riddle, Jane R., George H. Moeller, and William H. Smith, "Breaking New Ground in Urban America," *American Forests,* 83 (11):26, 1976.

Unsoeld, O., *National Organizations Involved in Urban Forestry,* USDA Forest Service, Southeastern Area, 1980.

USDA Forest Service, *The Principal Laws Relating to Forest Service Activities,* Agricultural Handbook No. 453, U.S. Government Printing Office, Washington, D.C., p. 83, 1974.

11
URBAN
FORESTRY ISSUES

Preliminary to a discussion of urban forestry issues is the understanding that issues change rapidly—they no longer remain as important, either by change or correction, and/or replacement with other issues. A case in point of perhaps both is energy. In our original text, written in 1977, we stated the following:

> *Of the issues facing urban forestry perhaps none is more important than energy. Scarcity and high costs of fossil fuels have focused attention on the utilization of wood from the urban forest for heating. There is little doubt that this trend will continue. Manufacturers of wood-burning stoves, furnaces, and fireplaces are currently hard-pressed to keep up with demand, and retail stores for heating equipment have opened in many cities. In addition to accelerating utilization of "waste" wood from the urban forest, there is increasing interest in the establishment of energy forests. Many of these are plantings on vacant lots or small acreages in urban vicinities to be managed and harvested for fuelwood.*
>
> *There is increasing emphasis on the use of trees and other plants to conserve energy. Trees for wind barriers and shrubs for insulation are becoming more important to the landscape. More consideration is being given to precise locations of deciduous trees to provide summer shade yet allow incoming solar radiation in winter. An interesting application of this concept is the use of dedicuous vines on passive solar collector walls to prevent excessive heat generation in summer.*
>
> *Changing energy sources have long-term implications for the urban forest. There is certain to be changes in our automobile-oriented society. It is conceivable that we will experience a vast reduction in automobile numbers as gasoline prices continue to increase. What is more likely, however, is a general conversion to smaller, more energy-efficient autos. In either situation, the design of urban areas will be influenced—perhaps the results will be narrower streets, more neighborhood shopping centers, parks, and other facilities. And this will certainly influence the composition and distribution of the urban forest.*

While energy has not faded completely as an urban forestry issue, it is no longer deemed as important, because much has been done to adjust to it and because, as of this writing, foreign oil prices have been greatly reduced. It is certain to emerge again as a major issue, however, simply because fossil fuel sources are finite.

Shortly after our original text writing (had we been more perceptive then, we would have detected it), the conservative movement began to manifest itself—first with the passage of Proposition 13 in California, and now well into the second term of the Reagan Administration—with traumatic impacts on public service budgets at all levels of government. This appears to be, without question, the major urban forestry issue of the decade. History must judge the wisdom of the movement, whether its promised long-term benefits will be greater than current costs. Forced by reduced budgets to make difficult choices, government administrators have often given lower priority to urban forestry programs, offering in many cases the rationale that "trees will grow by themselves anyway." In extreme cases, urban forestry programs have been eliminated. More commonly, programs have been altered, with functions being prioritized and some activities either discontinued or transferred. Activity transfer has in some cases been to volunteer organizations—perhaps a healthy result of the budget dilemma. Overall, though, there is little doubt that urban forestry has not benefited.

Reduced budgets for public urban forestry programs have partially settled the issue of public vs. private involvement in urban forestry—there is simply less public involvement. However, no evidence exists that the private sector has rushed in to fill the void, in spite of the often voiced assertion that government agencies were competing unfairly. There are, in fact, occasional lamentations of the loss of public education programs in urban forestry, perhaps in recognition of the "free" advertising value of such programs. This philosophy will perhaps contribute to future involvement by public agencies, particularly state and federal government, in urban forestry as service-type activities are deemphasized in favor of education programs.

Concerns for environmental protection will continue to influence urban forestry. Of direct concern are materials and equipment such as pesticides and noisy machinery that may adversely affect the environment. Also of direct concern are government regulations on the use of these items. Of less direct concern are the environmental factors that will influence the future of urban areas. Many of these are related to changing energy sources for heating, cooling, and transportation. Others are related to society's desires for a "green" environment of grass, trees, and open spaces.

Recent developments indicate the importance of the environmental issue. A veritable storm raged within the Congress and the Reagan Administration concerning the Environmental Protection Agency. The direct issue concerned the administration of a "super fund" for cleanup of toxic waste disposal sites and the alleged capitulation of the agency to industry's viewpoint concerning the chemical dioxin. The larger issue, however, is the general environmental philosophy (and policy) of the Reagan Administration. We offer the speculation that it is the handling of this issue, more than any other, that will have the strongest impact on the future of the nation.

As suggested in the preface, acceptance of urban forestry within forestry and related professions is a continuing issue. Within forestry, there are those who do not see it as "real forestry." In related professions, there are those who see it as a territorial invasion. While it is unrealistic to believe that urban forestry will ever be totally accepted by everyone, there are indications that progress is being made. For this progress to continue, three things appear necessary: (1) urban forestry must truly bring a workable systems approach to forest management to the urban environment; (2) urban forestry must provide a recognizable link for urban residents to "natural" forests; and (3) the public must recognize urban forestry's contributions.

Technology will continue to provide for new issues. Computer technology will result in the greatly increased efficiency and effectiveness of urban tree care programs, going far beyond current inventory systems to every aspect of program planning, implementation, and evaluation. Robotics will be applied to tree pruning and other cultural practices. New chemicals and systems for their application will be developed. Acceptability of all new technology must, however, depend on economic feasibility and social acceptance. It is the latter, social acceptance, or lack thereof, that creates issues.

Urban forestry will continue to be influenced by many issues in our society. What is perhaps more important is the influence that urban forestry itself will have. The well-being of and a better life for urban society are at issue. Through proper manipulation of the components of the urban forest—vegetation, water, soil, air, and animals (including people)—urban forestry can make a significant contribution to a better quality of life.

APPENDICES

APPENDIX 1

Schools for Urban Forestry in North America

University	Current				Planned				Syllabic Materials			Joint Prgms.		Plans Avail.	UF Cognates
	Courses		Curric.		Courses		Curric.		Curric.	CD	CO^b	Curr.	Plan^c		
	UG	G	UG	G^a	UG	G	UG	G							
Arkansas	−	−	−	−	−	−	−	−	−	−	−	−	−	−	+
California (Berkeley)	−	−	−	−	−	+	−	+	−	−	−	−	+	+	+
California Polytechnic (S.L.O.)	+	−	−	−	−	−	−	−	−	+	+	−	−	−	−
Colorado State	−	−	−	+	−	−	−	−	−	−	−	+	−	−	+
Duke	−	−	+	+	−	−	−	−	−	−	−	+	−	−	+
Florida	−	−	+	+	−	−	−	−	+	−	−	−	−	−	+
Georgia	+	−	−	−	−	−	−	−	−	+	+	−	−	−	−
Illinois (Urbana)	+	−	+	−	−	−	−	−	+	+	+	−	−	−	−
Western Illinois	−	−	+	−	−	−	−	−	+	+	+	−	−	−	−
Iowa State	+	−	−	−	−	−	−	−	−	−	−	−	−	−	−
Kansas State	+	−	+	−	−	+	−	+	−	+	+	−	+	+	+
Kentucky	−	−	−	−	−	−	−	−	−	−	−	−	−	+	+
Lakehead (Ontario)	+	−	−	−	−	−	−	−	−	+	+	−	+	−	+
Louisiana State	−	−	−	−	+	−	+	−	−	−	−	−	−	−	−
Louisiana Tech.	−	−	+	−	+	−	+	−	−	−	−	−	+	−	−
Maine	+	−	−	+	−	−	−	−	+	+	+	−	−	−	−
Michigan	+	+	+	+	−	−	−	−	+	+	+	−	−	−	+
Michigan State	+	+	+	−	−	−	−	−	+	+	+	−	−	−	−
Michigan Tech.	+	−	+	−	−	−	−	−	+	+	+	−	−	−	−
Minnesota	−	−	+	−	+	−	−	−	+	+	−	−	−	+	+
Missouri	−	−	−	−	+	−	−	−	−	−	−	−	−	−	+

	C1	C2	C3	C4	C5	C6	C7	C8	C9	C10	C11	C12	C13	C14	C15
Nebraska	+	—	—	—	—	—	—	—	—	+	+	—	—	—	—
New Hampshire	—	—	+	—	—	—	—	—	+	—	—	+	—	—	—
New York State (Syracuse)	+	+	+	+	+	+	+	—	+	+	+	—	—	—	—
North Carolina State	—	—	—	—	—	—	—	—	—	—	—	+	—	—	+
Ohio State	—	—	+	+	—	—	+	—	—	—	—	—	+	+	+
Oklahoma State	—	—	—	—	—	—	—	—	—	—	—	+	—	—	+
Oregon State	—	—	+	+	—	—	+	—	—	—	—	—	—	—	—
Pennsylvania State	+	+	+	—	—	—	—	—	+	+	+	+	—	—	+
Purdue	+	—	—	—	—	—	—	—	—	—	—	—	—	—	—
Rutgers	+	—	—	—	—	—	+	—	+	—	—	—	+	+	+
Stephen F. Austin	+	—	—	—	—	—	—	—	+	—	—	—	—	—	—
Tennessee	+	+	+	+	+	—	+	—	+	+	+	—	—	—	+
Texas A&M	+	+	+	+	+	+	+	—	+	+	+	—	+	+	+
Toronto	+	+	—	—	+	—	+	—	—	+	+	—	—	—	+
Vermont	+	—	—	—	—	—	—	—	—	+	+	—	—	—	—
Virginia Polytechnic	+	—	+	—	—	—	+	—	+	+	+	—	—	+	—
Washington	—	—	—	—	—	—	—	—	—	—	—	—	—	—	—
West Virginia	—	—	+	—	+	—	+	—	+	+	+	—	—	—	—
Wisconsin (Stevens Point)	+	—	—	—	—	—	—	—	—	—	—	—	—	—	+
Yale	—	+	—	+	—	—	—	—	+	—	—	+	+	+	+
TOTALS	19	7	18	10	9	3	8	2	17	19	19	7	7	7	20

[a]UG: undergraduate.
  G: graduate.
[b]Curric.: curriculum.
  CD: catalog description.
  CO: course outline.
[c]Curr.: current.
  Plan.: planned.
[d]Includes courses within and outside of forestry schools that include elements of urban vegetation management

Source: Andresen, John W., "North American Urban Forestry Educational Perspectives Including a Catalog of Urban Forestry and Arboricultural Courses," prepared for cooperative Forestry Branch, Forest Service, USDA, March 15, 1980.

APPENDIX 1a

Sample Urban Forestry Curriculum (University of Minnesota)

Urban foresters need a unique combination of skills and knowledge to manage forests in people-dominated environments. They must understand both the physical and the biological processes that affect tree growth and the role of trees in the ecosystem. They also must be able to plan urban forestry programs that meet the needs of residents, to promote these programs among decision makers and residents, and to manage the human and financial resources needed to accomplish urban forestry goals.

The College of Forestry at the University of Minnesota is one of only a few forestry schools in North America to offer a four-year curriculum in urban forestry. The curriculum gives students a foundation in basic and applied natural sciences. Part of this science background is gained during a three and a half week summer session at the University's Lake Itasca Forestry and Biological Station located at the headwaters of the Mississippi River. Students in the urban forestry curriculum also take communications, business, and social science courses to increase their managerial skills and give them an understanding of human needs and behavior.

A more specific list of required coursework in the urban forestry curriculum follows.

Subject Area	Quarter Credits	Subject Area	Quarter Credits
First Year		**Second Year**	
biology	10	horticulture	5
mathematics	10–14	landscape architecture	4
chemistry	10	soil science	4
physics	5	public speaking	4
communications	8	accounting	5
		economics	5
		computer programming	4
		statistics	
		social science	5
		business law	5
			4
Third Year		**Fourth Year**	
fisheries and wildlife	4	urban forest	6
horticulture	11	management,	
forest products	4	administration	

Subject Area	Quarter Credits	Subject Area	Quarter Credits
forest biology, silviculture, and pathology	14	nursery management	7
		forest recreation	3
forest meteorology, climatology	2	forest entomology	4
		forest-water relations	3
natural resource inventory	3	industrial relations or public relations	4
forest policy, economics	5	public project planning	4
aerial photo interpretation	3	technical writing	4

APPENDIX 2

Sample City Tree Ordinance for a Small Midwest Community

Be it ordained by the City Commission of the City of _____, Kansas:

Section 1: Definitions

Street trees: "Street trees" are herein defined as trees, shrubs, bushes, and all other woody vegetation on land lying between property lines on either side of all streets, avenues, or ways within the City.

Park Trees: "Park trees" are herein defined as trees, shrubs, bushes and all other woody vegetation in public parks having individual names, and all areas owned by the City, or to which the public has free access as a park.

Section 2: Creation and Establishment of a City Tree Board

There is hereby created and established a City Tree Board for the City of _____, Kansas, which shall consist of five members, citizens and residents of this city, who shall be appointed by the mayor with the approval of the Commission.

Section 3: Term of Office

The term of the five persons to be appointed by the mayor shall be three years except that the term of two of the members appointed to the first

265

board shall be for only one year and the term of two members of the first board shall be for two years. In the event that a vacancy shall occur during the term of any member, his or her successor shall be appointed for the unexpired portion of that term.

Section 4: Compensation

Members of the Board shall serve without compensation.

Section 5: Duties and Responsibilities

It shall be the responsibility of the Board to study, investigate, council, and develop and/or update annually, and administer a written plan for the care, preservation, pruning, planting, replanting, removal, or disposition of trees and shrubs in parks, along streets and in other public areas. Such plan will be presented annually to the City Commission and upon their acceptance and approval shall constitute the official comprehensive city tree plan for the City of _____, Kansas.

The Board, when requested by the City Commission, shall consider, investigate, make finding, report, and recommend upon any special matter of question coming within the scope of its work.

Section 6: Operation

The Board shall choose its own officers, make its own rules and regulations, and keep a journal of its proceedings. A majority of the members shall be a quorum for the transaction of business.

Section 7: Street Tree Species to be Planted

The list on the following page constitutes the official Street Tree species for _____, Kansas. No species other than those included in this list may be planted as Street Trees without written permission of the City Tree Board.

Section 8: Spacing

The spacing of Street Trees will be in accordance with the three species size classes listed in Section 7 of this ordinance, and no trees may be planted closer together than the following: Small Trees, 30 ft (9.1 m); Medium Trees, 40 ft (12.2 m); and Large Trees, 50 ft (15.2 m); except in special plantings designed or approved by a landscape architect.

Section 9: Distance from Curb and Sidewalk

The distance trees may be planted from curbs or curblines and sidewalks will be in accordance with the three species size classes listed in Section 7 of this ordinance, and no trees may be planted closer to any curb or sidewalk than the following: Small Trees, 2 ft (0.61 m): Medium Trees, 3 ft (0.91 m); and Large Trees, 4 ft (1.22 m).

Small Trees	Medium Trees	Large Trees
Apricot	Ash, green	Coffeetree, Kentucky
Crabapple, flowering (sp.)	Hackberry	Maple, silver
Goldenraintree	Honeylocust (thornless)	Maple, sugar
Hawthorn (sp.)	Linden or basswood (sp.)	Oak, bur
Pear, Bradford	Mulberry, red (fruitless, male)	Sycamore
Redbud	Oak, English	Sycamore, London planetree
Soapberry	Oak, red	Cottonwood (cottonless, male)
Lilac, Japanese tree	Pagodatree, Japanese	
Peach, flowering	Pecan	
Plum, purpleleaf	Birch, river	
Serviceberry	Osageorange (male, thornless)	
	Persimmon	
	Poplar, white	
	Sassafras	

Section 10: Distance from Street Corners and Fireplugs

No Street Tree shall be planted closer than 35 ft (10.67 m) to any street corner, measured from the point of nearest intersecting curbs or curblines. No Street Tree shall be planted closer than 10 ft (3.05 m) to any fireplug.

Section 11: Utilities

No Street Trees other than those species listed as Small Trees in Section 7 of this ordinance may be planted under or within 10 lateral ft (3.05 m) of any overhead utility wire, or over or within 5 lateral ft (1.52 m) of any underground water line, sewer line, transmission line, or other utility.

Section 12: Public Tree Care

The City shall have the right to plant, prune, maintain, and remove trees, plants, and shrubs within the lines of all streets, alleys, avenues, lanes, squares, and public grounds, as may be necessary to insure public safety or to preserve or enhance the symmetry and beauty of such public grounds. The City Tree Board may remove or cause or order to be removed any tree or part thereof which is in an unsafe condition or which by reason of its nature is injurious to sewers, electric power lines, gas lines, water lines, or other public improvements, or is affected with any injurious fungus, insect, or other pest. This Section does not prohibit the planting of Street Trees by adjacent property owners provided that the selection and location of said trees is in accordance with Sections 7 through 11 of this ordinance.

Section 13: Tree Topping

It shall be unlawful as a normal practice for any person, firm, or city department to top any Street Tree, Park Tree, or other tree on public property. Topping is defined as the severe cutting back of limbs to stubs larger than three inches in diameter within the tree's crown to such a degree so as to remove the normal canopy and disfigure the tree. Trees severely damaged by storms or other causes, or certain trees under utility wires or other obstructions where other pruning practices are impractical may be exempted from this ordinance at the determination of the City Tree Board.

Section 14: Pruning, Corner Clearance

Every owner of any tree overhanging any street or right-of-way within the City shall prune the branches so that such branches shall not obstruct the light from any street lamp or obstruct the view of any street intersection and so that there shall be a clear space of 8 ft (2.43 m) above the surface of the street or sidewalk. Said owners shall remove all dead, diseased, or dangerous trees, or broken or decayed limbs that constitute a menace to the safety of the public. The City shall have the right to prune any tree or shrub on private property when it interferes with the proper spread of light along the street from a street light, or interferes with the visibility of any traffic control device or sign.

Section 15: Dead or Diseased Tree Removal on Private Property

The City shall have the right to cause the removal of any dead or diseased trees on private property within the city, when such trees constitute a hazard to life and property, or harbor insects or disease that constitute a potential threat to other trees within the city. The City Tree Board will notify in writing the owners of such trees. Removal shall be done by said owners at their own expense within 60 days after the date of service of notice. In the event of failure of owners to comply with such provisions, the

City shall have the authority to remove such trees and charge the cost of removal on the owners property tax notice.

Section 16: Removal of Stumps

All stumps of street and park trees shall be removed below the surface of the ground so that the top of the stump shall not project above the surface of the ground.

Section 17: Interference with City Tree Board

It shall be unlawful for any person to prevent, delay, or interfere with the City Tree Board, or any of its agents, while engaging in and about the planting, cultivating, mulching, pruning, spraying, or removing of any Street Trees, Park Trees, or trees on private grounds, as authorized in this ordinance.

Section 18: Arborists License and Bond

It shall be unlawful for any person or firm to engage in the business or occupation of pruning, treating, or removing street or park trees within the City without first applying for and procuring a license. The license fee shall be $__ annually in advance; provided, however, that no license shall be required of any public service company or City employee doing such work in the pursuit of their public service company or City employee doing such work in the pursuit of their public service endeavors. Before any license shall be issued, each applicant shall first file evidence of possession of liability insurance in the minimum amounts of $__ for bodily injury and $__ property damage, indemnifying the City or any person injured or damaged resulting from the pursuit of such endeavors as herein described.

Section 19: Review by City Commission

The City Commission shall have the right to review the conduct, acts, and decisions of the City Tree Board. Any person may appeal from any ruling or order of the City Tree Board to the City Commission who may hear the matter and make a final decision.

Section 20: Penalty

Any person violating any provision of this ordinance shall be, upon conviction or a plea of guilty, subject to a fine not to exceed $_____ .

APPENDIX 2a

Model Municipal Tree Ordinance for the Atlanta, Georgia Area

Regulating the Protection, Maintenance, Removal, and
Planting of Trees on Certain Areas in the

City of _____

county of _____

state of ____ Georgia _____

ordinance No. _____

An ordinance regulating the protection, maintenance, removal, and planting of trees in the public streets, parks, other public places, and tree protection zones on designated private property under development; establishing a tree commission and establishing the office of a municipal arborist, as the agencies prescribing regulations relating to the protection, maintenance, removal, and planting of trees in the above-mentioned places; providing for the maintenance or removal of trees on private property when the public safety is endangered; and prescribing penalties for violations of its provisions.

Except for Sections 12 and 13, this ordinance in no way regulates protection, maintenance, removal, or planting of trees on the property of a one or two-family dwelling, where the owner of the property resides thereon.

Be it ordained by the Council of the Municipality of _____ , County of _____, State of Georgia.

Section 1: Value of Urban Trees

_____, Georgia, is situated in an area covered with a wide variety of trees and shrubs that are a vital part of the heritage passed to us by nature and our forefathers. This vegetation creates such an impression on visitors that the Atlanta area is nationally recognized for the beauty of its plant life. However, there is much concern over the recent, indiscriminate destruction of trees in both public and private places.

Trees are recognized as a valued asset, providing a healthier and more beautiful environment in which to live. They provide oxygen, shade, esthetics, and a priceless psychological counterpoint to the manmade, urban setting. Trees aid in preventing erosion, siltation of streams and reservoirs, flash flooding, and air, noise, and visual pollution.

Trees are economically beneficial in attracting new industry, residents, and visitors. Healthy trees of the right size and species, growing in the right places enhance the value and marketability of property, and promote the

270

stability of desirable neighborhoods, thus helping to prevent the emergence of blighted areas and slum conditions.

Section 2: Short Title

This Ordinance shall be known and may be cited as the Municipal Tree Ordinance of the Municipality of _____, County of _____, State of Georgia.

Section 3: Definitions

For the purpose of this Ordinance the following terms, phrases, words, and their derivations shall have the meaning given herein. When not inconsistent with the context, words used in the present tense include the future, words in the plural include the singular, and words in the singular include the plural. The word *SHALL* is mandatory and not merely directory.

A. *Municipality* is the municipality, city, town, village, subdivision, or otherwise designated political jurisdiction of _____, County of _____, State of Georgia.

B. *Council* is the council, board, commission, or otherwise designated governing body of the municipality.

C. *Park and Street Tree Department* is the department of "Parks and Street Trees," "Parks and Forestry," "Forestry," "Street Trees," or other designated department of the Municipality under whose jurisdiction park and/or street trees fall.

D. *Municipal Arborist* is the Municipal Arborist, Forester, Horticulturist, Landscape Architect, or other qualified designated official of the Municipality assigned to carry out the enforcement of this Ordinance.

E. *Person* is any person, firm, partnership, association, corporation, company, or organization of any kind, including public utility and municipal department.

F. *Street or Highway* is the entire width of every public way or right-of-way when any part thereof is open to the use of the public, as a matter of right, for purposes of vehicular and pedestrian traffic.

G. *Park* shall include any public parks having an individual name.

H. *Public Place* shall include any other ground owned by the Municipality.

I. *Property Line* shall mean the outer edge of a street or highway.

J. *Treelawn* is that part of a street or highway, not covered by sidewalk or other paving, lying between the property line and that portion of the street or highway usually used for vehicular traffic.

K. *Tree:*

(1) Any living pine that has a well-defined stem with a caliper of at least 6 in. (15.24 cm), at 4.5 ft (1.4 m) from the ground level.

(2) Any living, self-supporting nonpine (except in 3 below) that has a well-defined stem with a caliper of at least 3 in. (7.62 cm), at 4.5 ft (1.4 m) from the ground level.

(3) Any dogwood, redbud, or other conspicuously flowing tree, as listed by the Municipal Arborist, which has a well-defined stem of at least 2 in. (5.1 cm), at 4.5 ft (1.4 m) from the ground level.

L. *Public Trees* shall include all trees now or hereafter growing on any street, park, or any other public place.

M. *Property Owner* shall mean the person owning such property as shown by the County Plan of _____ County, State of Georgia.

N. *Tree Removal* includes any act that will cause a tree to die within a three-year period.

Section 4: Establishment of a Tree Commission

A. There shall be created a commission to be known and designated as the "Tree Commission" composed of nine (9) citizens of the Municipality of _____, a majority of whom shall be residents of the Municipality of _____, or of _____ County. Six (6) of said members shall be appointed by the Mayor with approval of the Council and three (3) of these shall be persons professionally trained in fields such as forestry, botany, horticulture, and landscape architecture. The seventh (7th) member shall be the Director of Public Service who shall be an ex-officio member, the eighth (8th) member shall be the Superintendent of the Department of Parks and Street Trees who shall be an ex-officio member, and the ninth (9th) member shall be the Municipal Arborist who shall be an ex-officio member. All members of the Commission shall serve without pay. The six (6) members appointed by the Mayor shall be appointed as follows: two (2) for two (2) years; two (2) for three (3) years; and two (2) for four (4) years, and serve until their successors are duly appointed and approved by the Council. Successors to those members appointed by the Mayor shall, thereafter be appointed for terms of four (4) years. Vacancies caused by death, resignation, or otherwise shall be filled for the unexpired term in the same manner as original appointments are made.

B. The duties of said "Tree Commission" shall be as follows:

To study the problems and determine the needs of the Municipality in connection with its tree planting program.

To recommend to the Municipal Arborist, the type and kind of trees to be planted upon such municipal streets, parks, public places, and other tree protection zones.

To assist the properly constituted officials of the Municipality, as well as the Council and citizens of the Municipality, in the dissemination of news and information regarding the protection, maintenance, removal, and planting of trees on public or private property, and to make such recommendations from time to time to the Municipal Council as to desirable legislation concerning the tree program and activities for the Municipality.

To provide monthly and special meetings at which the subject of trees may be discussed by at least three members of the Commission, officers, and personnel of the Municipality, and all other persons interested in the tree program.

To hear appeals as set forth in Section 19.

C. That within a reasonable time after the appointment of said Commission and the approval of the members thereof, upon call of the Mayor, said Commission shall meet and organize by the election of a chairman and the appointment of the Municipal Arborist as secretary. The said Commission shall then provide for the adoption of rules and procedures and for the holding of regular and special meetings as said Commission shall deem advisable and necessary in order to perform the duties set forth.

Section 5: Appointment and Qualifications of the Municipal Arborist

The Municipal Arborist shall, where possible, be appointed from a Merit System roster established by competitive examination after a personal interview, or where a Merit System does not exist, by a competitive examination and interview given by the Tree Commission of the Municipality. Upon satisfactory completion of a six (6) months probationary period he or she shall hold office as long as he satisfactorily performs the duties of his office. He shall be a person skilled and trained in the arts and sciences of municipal arboriculture, and shall hold a college degree or its equivalent in arboriculture, ornamental, or landscape horticulture, forestry, or other closely related field. If and when there is a State (Forester, Arborist, Horticulturist, etc.) Examining Board, he shall have passed the state examination. He shall have had at least three (3) years experience in municipal shade tree work or its equivalent.

The position of Municipal Arborist shall be funded and filled within one year of the adoption of this Ordinance.

Section 6: Salary of the Municipal Arborist

The Municipal Arborist shall receive a salary commensurate with his training and experience, as full compensation for all services rendered and in lieu of all fees.

273

Section 7: Duties of the Municipal Arborist

The Municipal Arborist shall affirm the rules and regulations of the *Arboricultural Specifications and Standards of Practice* (appended to this Ordinance) governing the protection, maintenance, removal, and planting of trees on the streets, parks, public places, and other Tree Protection Zones in the Municipality. He shall direct, regulate, and control the protection, maintenance, removal, and planting of all trees growing now or hereafter in the streets, parks, public places, and other tree protection zones of the Municipality. He is also charged with planning, constructing, and maintaining a plant nursery/tree bank for the purpose of supplying trees and shrubs for planting on public land. He is further charged with keeping informed of environmental and technical changes that could affect the trees of the Municipality. He shall cause the provision of this Ordinance to be enforced. In his absence these duties shall be the responsibility of a qualified alternate designated by the Municipal Arborist.

Section 8: Authority of the Municipal Arborist

A. Jurisdiction: The Municipal Arborist shall have the authority and jurisdiction of regulating the protection, maintenance, removal, and planting of trees on streets, parks, public places, and other tree protection zones.

B. Enforcement: The Municipal Arborist shall be vested with police power for the purpose of enforcing this Ordinance and to insure that provisions of this Ordinance are not violated including, but not limited to, the issuance of citations for the violation of any provisions of this Ordinance.

C. Supervision: The Municipal Arborist shall have the authority and it shall be his duty to supervise or inspect all work done under a permit issued in accordance with the terms of this Ordinance.

D. Condition of Permit: The Municipal Arborist shall have the authority to affix reasonable conditions to the granting of a permit in accordance with the terms of this Ordinance.

E. Inventory of Existing Trees: The Municipal Arborist shall have the responsibility of inventorying (and classifying as to location, species, size, condition, and evaluation) the existing trees on streets, parks, and other public places, as an integral part of the Master Street Tree Plan. This Inventory shall be periodically updated.

F. Master Street Tree Plan: The Municipal Arborist shall also have the authority to formulate a Master Street Tree Plan with the advice and approval of the Tree Commission. The Master Street Tree Plan shall include the Inventory of Existing Trees, and shall specify the species of trees to be protected, maintained, removed, and/or planted on each of the streets, parks, and other public places of the Municipality. From and after the effective date of the Master Street Tree Plan, or any amendment thereof, all tree work shall conform thereto.

(1) The Municipal Arborist shall consider all existing and future utility and environmental factors when recommending the planting of a tree species or other tree work for each of the streets, parks, and other public places of the Municipality.

(2) The Municipal Arborist, with the approval of the Tree Commission, shall have the authority to amend or add to the Master Street Tree Plan at any time that circumstances make it advisable.

Section 9: Tree Protection Zone

The Tree Protection Zone shall:

A. Include all streets, parks, and other public places such as the grounds of public buildings, schools, libraries, etc.

B. Include all property planned for, or under development, redevelopment, razing, or renovating . . . unless the property is judged (based on current zoning) by the Municipal Arborist to be commercial forest land and the trees to be removed are to be utilized for some wood product.

C. Include that land fifteen (15) feet (4.57 m) beyond the roof line of any building.

D. Include all land within a horizontal distance of fifty (50) feet (15.24 m) from the right-of-way of any street or highway.

E. Include that portion of designated private property identified as the front, side, and rear yard, set-back requirements as established by the Municipal Zoning Code.

This section shall not be interpreted to prohibit the removal, as determined by the Municipal Arborist, of:

1. Dead, dying, diseased, insect infested, or hazardous trees.
2. Unwanted weed species.
3. The necessary minimal amount of trees and limbs by survey parties.

Nor shall this Section be interpreted to prohibit the protection, maintenance, removal, or planting of trees on the property of one- or two-family dwellings where the property owner resides thereon.

Section 10: Protection of Trees

Within each Tree Protection Zone:

A. During development, redevelopment, razing, or renovating no more

275

than 50 percent of the trees shall be cut, damaged, destroyed, or removed except by specified permit.

B. All trees within 30 ft (9.1 m) of any excavation or construction of any building, structure, or street work shall be guarded through the length of the project with a good substantial fence, frame, or box not less than 4 feet (1.2 m) high and 8 ft (2.4 m) square, or at a distance in feet from the tree equal to the diameter of the trunk in inches at 4.5 ft (1.4 m) from the ground, whichever is greater. All equipment, building materials, chemicals, dirt, or other debris shall be kept outside the barrier at all times.

C. No person shall excavate any ditches, tunnels, trenches, or lay any drive within a radius of 10 ft (3.05 m) from any tree without first obtaining a written permit from the Municipal Arborist.

D. No person shall intentionally damge, cut, carve, transplant, any tree; attach any rope, wire, nails, advertising posters, or other contrivance to any tree; allow any gaseous, liquid, chemical, or solid substance that is harmful to such trees to come in contact with them; or set fire or permit any fire to burn when such fire or the heat thereof will injure any portion of any tree without first obtaining a written permit from the Municipal Arborist.

E. No person shall deposit, place, store, or maintain any stone, brick, sand, fill dirt, concrete, or other materials that may impede the free passage of water, air, and fertilizer to the roots of any tree growing therein, except by written permit of the Municipal Arborist.

Section 11: Site Plan

The Site Plan for development or improvement of any tract or parcel of land shall be evaluated and approved by the Municipal Arborist before a building permit can be issued. The Municipal Arborist may inspect the site before approving or disapproving the site plan and at any time during the development, redevelopment, renovating, or razing thereafter.

The Site Plan, in addition to the usual requirements of the Zoning Code, shall show the following:

A. Approximate location, size, and species of all existing:

1. Trees to be maintained.
2. Trees to be removed.

B. Specifications for:

1. Protection of existing trees during the development, redevelopment, renovating, or razing.
2. Grade changes or other work within a tree's drip line.
3. Disposal of trees to be removed.

276

C. Any intended replanting, specifying the location, species, size, and a completion date for the seasonable planting of trees.

Section 12: Obstruction: Trees Pruned

It shall be the duty of any person or persons owning or occupying real property bordering on any street upon which property there may be trees, to prune such trees in such manner that they will not obstruct or shade the street lights, obstruct the passage of pedestrians on sidewalks, obstruct vision of traffic signs, or obstruct the view of any street or alley intersection. The minimum clearance of any overhanging portion thereof shall be 10 ft (3.05 m) over sidewalks, and 12 ft (3.66 m) over all streets except truck thoroughfares, which shall have a clearance of 16 ft (4.88 m).

A. Notice to Prune. Should any person or persons owning real property bordering on any street fail to prune trees as herein above provided, the Municipal Arborist shall order such person or persons, within three (3) days after receipt of written notice, to so prune such trees.

B. Order Required. The order required herein shall be served by certified mail to the last known address of the property owner.

C. Failure to Comply. When a person to whom an order is directed shall fail to comply within the specified time, it shall be lawful for the Municipality to prune such trees.

Section 13: Insects and Diseases

The Municipal Arborist shall have the authority to direct private property owners to treat or remove a tree suffering from a transmittable disease or insect infestation when such tree is on private property but may affect the health of trees on public property. The owner of said tree shall be notified of the condition and recommended treatment in writing and shall be given a period of one (1) week to instigate the treatment or removal. If not treated or removed within a period of one (1) additional week, the Municipal Arborist shall cause the treatment or removal to be carried out and the cost may be assessed to the tree owner.

Section 14: Emergencies

In the case of officially declared emergencies, such as windstorms, ice storms, or other disasters, the requirements of Sections 10 and 16 shall be waived so that the requirements of this Ordinance would in no way hinder private or public work to restore order in the Municipality. This work shall follow maintenance standards as outlined by the Municipal Arborist.

277

Section 15: Moving Large Objects

No person shall move any building or other large objects that may injure public trees, or parts thereof, without first having obtained the approval of the Municipal Arborist.

Section 16: Permits Required

A. Planting, Maintenance, or Removal

(1) Within any tree protection zone, during development, redevelopment, razing, or renovating, no person shall spray, prune, cut above ground, remove or otherwise disturb more than 50 percent of the tree without first filling an application and procuring a permit from the Municipal Arborist.

(2) No person shall plant, spray, fertilize, prune, cut above ground, remove, or otherwise disturb any tree on any street, park, other public place, or other tree protection zone without first filling an application and procuring a permit from the Municipal Arborist.

(3) The person receiving the permit shall abide by the *Arboricultural Specifications and Standards of Practice*.

(4) Application for permits must be made at the office of the Municipal Arborist not less than seventy-two (72) hours in advance of the time the work is to be done.

(5) The Municipal Arborist shall issue the permit provided for herein if, in his or her judgment, the proposed work is desirable and the proposed method and workmanship to be used are of a satisfactory nature. Any permit granted shall contain a definite date of expiration and the work shall be completed in the time allowed on the permit and in the manner as therein described. Any permit shall be void if its terms are violated.

(6) Notice of completion shall be given to the Municipal Arborist for his inspection within five (5) days.

B. Planting

(1) The application required herein shall state the number of trees to be set out or planted in public places, the location, grade, species, cultivar or variety of each tree, the method of planting, and such other information as the Municipal Arborist shall find reasonably necessary to a fair determination of whether a permit should be issued.

(2) Whenever any tree shall be planted or set out in conflict with the provisions of this section, it shall be lawful for the Municipal Arborist to remove or cause removal of the same.

C. Maintenance

The application required herein shall state the number and kinds of trees to be sprayed, fertilized, pruned, or otherwise preserved; the kind of treatment to be administered; the composition of the spray material to be applied; and such other information as the Municipal Arborist shall find reasonably necessary to a fair determination of whether a permit should be issued.

D. Removal, Replanting, and Replacement

1. Wherever it is necessary to remove a tree from a treelawn in connection with the paving of a sidewalk, or the paving or widening of the portion of a street or highway, the Municipality shall replant the tree or replace it. Provided that conditions prevent planting on treelawns, this requirement will be satisfied if any equivalent number of trees of the same size and species as provided in the *Arboricultural Specifications and Standards of Practice* are planted in an attractive manner on the adjoinint property.

2. No person shall remove a tree from any tree protection zone for the purpose of construction, or for any other reason, without first filing an application and procuring a permit from the Municipal Arborist, and without replacing the removed tree in accordance with the adopted *Arboricultural Specifications and Standards of Practice*. Such replacement shall meet the standards of size, species, and placement as provided for in a permit issued by the Municipal Arborist. The person or property owner shall bear the cost of removal and replacement of all trees removed.

E. Permit Fees

A fee of _____ dollars ($_____) shall be assessed for each permit and shall be applicable to the particular job as specified by the permit. The Municipality and public utilities companies shall be exempt from acquiring individual job permits to perform necessary tree maintenance on public rights-of-way or public lands.

All permits issued for the installation or maintenance of public utilities or public works that affect trees shall be certified annually (semiannually) by the Municipal Arborist. The work of trimming or other operations affecting public trees shall be limited to the actual necessities of the service of the public utility or Municipal Department and shall be performed according to the *Arboricultural Specifications and Standards of Practice*. Permits shall expire one (1) year (6 months) from date of issuance.

Section 17: Registration of Tree Experts

To protect the public, the Municipality shall require any person who removes or maintains trees as a business on public or private lands to be:

A. Examined by the Tree Commission on his or her ability to carry out such work before he is granted a specific license (in addition to the normal business license) to practice as a tree expert. The examination shall consist of a standardized field test and written test. Supervisors of field crews shall be examined in the field on their ability to property treat and remove trees. When a statewide licensing law for tree experts exists, the above examination and specific license might not be required.

B. Bonded.

C. Insured (medical and liability).

A nonrefundable _____dollar ($_____) fee shall accompany each application for examination.

Employees of the Municipality responsible for maintaining public trees shall be examined in the field but shall not be subject to registration as a tree expert. They shall, however, be required to attend at least one day of training per year. Such training shall be sponsored by the Municipal Arborist and shall consist of up-to-date information and techniques on the planting, care, removal, and insects and diseases of trees.

Section 18: Nurseries and Botanical Gardens

All state-approved and governmental plant and tree nurseries and botanical gardens shall be exampt from the terms and provisions of this Ordinance only in relation to those trees that are so planted and growing for the sale or intended sale to the general public in the ordinary course of business or for some public purpose.

Section 19: Interference with Municipal Arborist

No person shall hinder, prevent, delay, or interfere with the Municipal Arborist or any of his assistants while engaged in carrying out the execution or enforcement of this Ordinance; provided, however, that nothing herein shall be cqnstrued as an attempt to prohibit the pursuit of any remedy, legal or equitable, in any court or competent jurisdiction for the protection of property rights by the owner of any property within the Municipality.

Section 20: Appeals

Any adjustment of the standards required by this Ordinance or an appeal of a decision of the Municipal Arborist shall be taken to the Tree Commission. The Tree Commission, upon receipt of such request, on forms provided by the Municipal Arborist, shall have the authority and duty to consider and act upon the request. This application shall clearly and in detail state what adjustments or requirements are being requested, reasons such adjustments are warranted, and shall be accompanied with such supplementary data as is deemed necessary to substantiate the adjustment. The Tree Commission may approve, modify, or deny the requested adjustment, based upon protection of public interest, preservation of the intent of this Ordinance, and possible unreasonable and unnecessary hardships involved in the case.

The Tree Commission shall act on the application as expeditiously as possible and shall notify the applicant in writing within five (5) days of the action taken.

Within fourteen (14) days after notification of the decision, but not thereafter, any decision of the Tree Commission may be appealed by the applicant to the Council. This appeal shall be on a form provided by and filed with the Municipal Arborist and the applicant shall be notified of the time and place this appeal will be heard.

The Council shall act on the application as expeditiously as possible and shall notify the applicant in writing within five (5) days of the action taken.

Section 21: Violation and Penalty

Any person violating or failing to comply with any of the provisions of this Ordinance shall be subject to court action and may be found guilty of a misdemeanor, and upon conviction thereof, shall be fined a sum not exceeding $250.00 or may be imprisoned for a term not exceeding ninety (90) days, or both.

In addition, the person shall replace the tree with a size and species recommended by the Municipal Arborist.

Each tree affected by noncompliance with this resolution shall constitute a separate violation.

Violation of this Ordinance shall be the basis of withholding a final inspection permit until such violation is corrected to the satisfaction of the Municipal Arborist, Tree Commission, Council or courts.

Section 22: Legality of Ordinance and Parts Thereof

Should any section, clause, or provisions of this Ordinance be declared by the Courts to be invalid, the same shall not affect the validity of the Ordinance as a whole, or parts thereof, other than the part so declared to be invalid.

All other Ordinances pertaining to shade trees are hereby repealed.

Section 23: Effective Date

This Ordinance is hereby declared to be of immediate necessity for the preservation of public peace, health, and safety, and shall be in full force and effective from and after its passage and publication as provided by law.

Passed this _____

day of _____

19 _____

Signed this _____

day of _____

19 _____

_____

Mayor

Attest _____

Municipal Clerk

Appendix 2b

Arboricultural Specifications and Standards of Practice

The Municipal Arborist shall have the authority to affirm the rules and regulations of the *Arboricultural Specifications and Standards of Practice* governing the planting, maintenance, removal, fertilization, pruning, bracing, and spraying of trees in the streets, parks, other public places, and tree protection zones in the Municipality.

A. Policy

(1) All work on protected trees shall comply with the "Municipal Tree Ordinance" of the Municipality of _____, County of _____, State of Georgia.

(2) The *Arboricultural Specifications and Standards of Practice* shall be adhered to at all times, but may be amended whenever experience, new research, or laws indicate improved methods, or whenever circumstances make it advisable, with the approval of the Tree Commission.

(3) The policy of the Department of Parks and Street Trees shall be to cooperate with the public, property owners, other Municipal departments, and appropriate not-for-profit organizations at all times.

(4) No trees shall be removed from any tree protection zone unless they

constitute a hazard to life or property, a public nuisance, or because a revision of planting plans necessitates.

B. Species, Cultivars, or Varieties

(1) The Municipal Arborist shall prepare lists of trees acceptable for planting in tree protection zones of the Municipality. Undesirable trees shall not be recommended for general planting and their use, if any, shall be restricted to special locations where, because of certain characteristics of adaptability or landscape effect, they can be used to advantage.

(2) Only desirable, long-lived trees of good appearance, beauty, and adaptability that are generally free from injurious insects or disease shall be planted in tree protection zones. The Tree Commission, in conjunction with the Municipal Arborist, shall review at least once every two (2) years the species, cultivars, and varieties included on the approved list to determine if any should be removed for any reason or if certain new species, cultivars, or varieties of proven dependability and value should be added.

(3) Where street blocks have been assigned a particular species or variety on the Master Street Tree Plan, only these shall be planted subject to revision by the Municipal Arborist and approved by the Tree Commission.

C. Planting

(1) Size

a. Unless otherwise specified by the Municipal Arborist, all medium to large deciduous tree species and their cultivars and varieties shall conform to American Association of Nurserymen Standards and be at least 1¼ to 1½ in. (3.17 to 3.81 cm) in diameter, 6 in. (15.24 cm) above ground level, and at least 8 to 10 ft (2.44 to 3.05 m) in height when planted. The crown shall be in good balance with the trunk.

b. All small deciduous tree species and their cultivars or varieties, shall be at least 5 to 6 ft (1.52 to 1.83 m) or more in height and have six (6) or more branches.

(2) Grade

Unless otherwise allowed for specific reasons, all trees shall have comparatively straight trunks, well-developed leaders and tops, and roots characteristic of the species, cultivar, or variety and shall show evidence of proper nursery pruning. All trees must be free of insects, diseases, mechanical injuries, and other objectionable features at the time of planting.

(3) Location and Spacing

a. Based on a 40-year cycle, no tree that will attain a trunk diameter greater than 12 to 15 in. (30.48 to 38.1 cm) shall be planted in a treelawn less than 3 to 5 ft (0.91 to 1.52 m) in width. In treelawns less than 3 ft (0.91

m) in width, or where overhead lines or building setback presents a special problem, the selection of site and species shall be determined by the Municipal Arborist.

b. Where there is a treelawn less than 3 ft (0.91 m) in width, legal steps should be taken to obtain easement rights to plant beyond the sidewalk on private property. Such easements should contain provisions that grant the Municipality permission to select, plant, maintain, and remove such trees under the direction of the Municipal Arborist.

c. Trees shall be planted at least 30 ft (9.14 m) from street intersections and at least 15 ft (4.54 m) from driveways and alleys.

d. No tree shall be planted closer than 10 ft (3.05 m) to a utility pole.

e. Spacing of trees should be determined by the Municipal Arborist according to local conditions, the species, cultivars, or varieties used and their mature height, spread, and form. Generally, all large trees shall be planted 40 to 60 ft (12.19 to 18.29 m) on center; all medium-sized trees shall be planted a minimum of 35 ft (10.64 m) on center; and all small trees shall be planted a minimum of 25 ft (7.62 m) on center.

f. All planting on unpaved streets without curbs must have the special permission of the Municipal Arborist who shall determine the tree's location so it will not be injured or destroyed when the street is curbed and paved.

(4) Methods of Planting and Support

a. Most small deciduous trees may be moved bare-rooted unless otherwise indicated. Roots of bare-rooted trees should be protected against drying out.

b. All coniferous trees shall be moved balled and burlapped. Balled roots should be prevented from drying out at the surface of the ball and protected against injurious freezing.

c. Pits dug for planting of bare-root plants shall be a minimum of 12 in. (30.48 cm) larger in diameter than the root system so as to be of sufficient size to accommodate the roots without crowding. For balled trees, the pits shall be a minimum of 12 in. (30.48 cm) larger in diameter than the diameter of the ball of soil to allow proper backfill.

d. Plants shall be planted no deeper than previously grown, with due allowance for settling.

e. In poorly drained soil, artificial drainage shall be provided to properly drain the soil about the plant roots or tolerant species selected.

f. Acceptable top soil, compost, peat moss, or other acceptable soil mixtures shall be placed about the roots, or in the backfill around the ball. When the planting is completed, the entire root area shall be thoroughly saturated with water.

g. Excessive pruning at the time of transplanting should be avoided.

The extent of top pruning should be based on the ability of the plant roots to function.

h. Trees shall be suitably wrapped and guyed, or supported in an upright position, according to accepted arboricultural practices. The guys or supports shall be fastened so that they will not girdle or cause serious injury to the tree or endanger public safety.

D. Early Maintenance

(1) General
Newly planted trees require special attention to maintenance practices during one or two growing seasons following planting. All maintenance practices shall follow approved arboricultural standards.

(2) Watering
Ample soil moisture shall be maintained following planting. A thorough watering each five (5) to ten (10) days, depending on soil type and drainage provisions, is usually adequate during the growing season. A soil auger or sampling tube is used to check the adequacy of moisture in the soil ball and/ or backfill.

(3) Fertilization
Provision of good drainage and adequate moisture of the prepared back-fill, and the soil ball of balled plants, is more important than fertilization immediately following planting. However, adequate quantities of the essential nutrient elements should be available after new growth starts.

(4) Insect and Disease Control
Measures for the control of insects and diseases shall be taken as shown necessary by frequent and thorough inspections. Plants in a weakened condition following transplanting are often more susceptible to insects, especially borers, and some diseases than are vigorously growing trees. Where it is necessary to spray, insecticides or fungicides shall be used that are recommended for safe and effective control.

(5) Pruning
a. Pruning practices to be followed in the first few years after planting shall consist of removing dead, broken, or injured branches, the suppression of rank, uneven growth, and usually the removal of water sprouts. Feather growth shall be removed when it reaches pencil size in diameter.

b. Pruning shall be practiced subsequent to transplanting and as necessary thereafter to assure sturdy crotch development.

c. Tree heads shall be raised as growth characteristics and location dictates. Newly planted trees need not have lower branches removed until they are well established. Eventually, trees should have the lower branches removed to a height of at least seven (7) feet, unless they are located in areas where the lower branches do not impede traffic.

E. General Maintenance

(1) Pruning and Removal

a. No topping or dehorning of trees shall be permitted except by written permission of the Municipal Arborist. Proper cabling and bracing shall be substituted for this practice wherever possible.

b. All large, established trees shall be pruned to sufficient height to allow free passage of pedestrians and vehicular traffic; 10 ft (3.05 m) over sidewalks and 12 ft (3.66 m) over all streets except those that are subject to truck traffic, which shall have a clearance of 16 ft (4.88 m).

c. It shall be the policy of the Municipal Arborist to cooperate with the Municipal or Utility Lighting Engineer, and vice-versa, in the placement and height of lighting standards and in the development of a system of tree pruning to give effective street illumination.

d. All cuts shall be made with a saw or pruner and only at the nodes or crotches. No stubs shall be left. No spurs or climbing irons shall be used in the trees, except when trees are to be removed.

e. All dead, crossed, and rubbing branches shall be removed.

f. All wounds over 2 in. (5.08 cm) in diameter shall be treated with a suitable tree wound dressing.

g. All tools being used on a tree suspected to be infected with a contagious disease shall be disinfected before being used on another tree.

h. Whenever streets are to be blocked off to public service, police and fire departments shall be notified of the location and length of time that the street will be blocked. Notification shall be given to these departments upon the removal of such barriers or if such barriers are to remain longer than originally expected.

i. To protect the public from danger, suitable street and sidewalk barriers, highway cones, or signs shall be placed on all barriers or obstructions remaining in the street after dark.

j. The stumps of all removed trees shall be cut to at least 3 in. (7.62 cm) below the ground, and soil shall be replaced and the area leveled. If the area where the tree is removed is to be paved, the tree should be cut or stump removed at least 6 in. (15.24 cm) below the ground.

(2) Spraying

a. Suitable precautions shall be taken to protect and warn the public that spraying is being done.

b. Spraying shall be done only for the control of specific diseases or insects, with the proper materials in the necessary strength, and applied at the proper time to obtain the desired control. All spraying practices shall conform to federal and state regulations.

c. Dormant oil sprays shall not be applied to sugar maple, Japanese maple, beech, flowering dogwood, hickory, walnut, and most crabapple trees. Dormant oil sprays shall be applied to other trees only when the air temperature is 40°F (5°C) or above and when it is not likely to drop below this temperature for a period of twenty-four (24) hours.

(3) Fertilization

a. Fertilization of public trees shall follow the National Arborist Association or other accepted arboricultural standards.

b. Formulations, rates, and methods of application of fertilizers shall be specified by the Municipal Arborist.

(4) Cavities

Extensive cavity work should be performed on trees only if they are sufficiently high in value to justify the cost. All cavity work shall conform to the National Arborists Association or other accepted arboricultural standards.

(5) Cabling and Bracing

a. As a general rule, cables should be placed approximately two-thirds (2/3) of the distance between the crotch and top branch ends. Rust-resistant cables, thimbles, and lags should be used. The ends of a cable should be attached to hooks or eyes of lags or bolts, and thimbles must be used in the eye splice in each end of the cable. Under no circumstance shall cable be wrapped around a branch.

b. All cabling and bracing practices with screw rods shall follow National Arborists Association or other accepted arboricultural standards.

F. Amend

The Municipal Arborist shall have the authority to modify, amend, or extend, with the approval of the Tree Commission, the *Arboricultural Specifications and Standards of Practice* at any time that experience indicates improved methods or whenever circumstances make it advisable.

APPENDIX 3

A City Street Tree Inventory System

	Instructions
City and Date:	Fill in on first tally sheet of the day.
Street:	Use a street name.
Block:	Use map code or block number. Blocks are always inventoried from south to north and west to east.
Side:	Street side—east, west, north, or south.
Species:	Use species codes on cards. Record vacant spots as "Blank."
Diameter:	DBH—to nearest even numbered inch.
Condition:	Use codes 1 through 4.

Code	
1 (good)	Healthy vigorous tree. No apparent signs of insect, disease, or mechanical injury. Little or no corrective work required. Form representative of species.
2 (fair)	Average condition and vigor for area. May need corrective pruning or repair. May lack desirable form characteristic of species. May show minor insect injury, disease, or physiological problem.
3 (poor)	General state of decline. May show severe mechanical, insect, or disease damage, but death not imminent. May require major repair or renovation.
4 (dead or dying)	Dead or death imminent from Dutch elm disease or other causes.

Management Needs: **Use codes 1 through 8.**

Code	
1	Minor pruning
2	Major pruning
3	Wound repair
4	Feeding (fertilizer, iron, or other elements)
5	Insect control
6	Disease control
7	Removal
8	None

Removal Difficulty:
*Code**

1	Easy
2	Medium
3	Difficult

Planting Needs: Base on vacant spots, and use 30 to 50 ft (9.14 to 15.24 m) spacing. Consider all restricting factors.

Code§

1	Small (mature height, 25 ft) (7.62 m)
2	Medium (mature height, 60 ft) (18.28 m)
3	Large (mature height, 90 ft) (27.43 m)

*Related to time and cost.   §Species codes could be used directly.

Species Codes: City Street Tree Inventory

001 Ash, green	024 Kentucky coffeetree	047 Pine, white
002 Ash, white	025 Locust, black	048 Poplar, Lombardy
003 Baldcypress	026 Magnolia	049 Poplar, white
004 Basswood	027 Maple, Norway	050 Purpleleaf plum
005 Birch (sp.)	028 Maple, red	051 Redbud
006 Boxelder	029 Maple, silver	052 Redcedar
007 Bradford pear	030 Maple, sugar	053 Smoke tree
008 Buckeye	031 Mountain ash	054 Soapberry
009 Catalpa	032 Mulberry, red	055 Spruce (sp.)
010 Cottonwood	033 Oak, bur	056 Sweetgum
011 Elm, American	034 Oak, chinquapin	057 Sycamore or planetree
012 Elm, hybrid	035 Oak, English	058 Tree of heaven
013 Elm, Siberian	036 Oak, pin	059 Tulip poplar
014 Fir (sp.)	037 Oak, red	060 Walnut, black
015 Flowering crab (var.)	038 Oak, shingle	061 Willow
016 Fruit (sp.)	039 Oak, white	062
017 Ginkgo	040 Osage orange	063
018 Goldenraintree	041 Pawpaw	064
019 Hackberry	042 Pecan	065
020 Hawthorn (sp.)	043 Persimmon	066
021 Hickory (sp.)	044 Pine, Austrian	067
022 Honeylocust	045 Pine, ponderosa	068
023 Japanese pagodatree	046 Pine, Scotch	069

Tally Sheet
City Street Tree Inventory

City <u>Anytown</u>   Date <u>12/1/76</u>

Street <u>Pine</u> Block <u>1</u> Side <u>W</u>   Street <u>Pine</u> Block <u>1</u> Side <u>E</u>

Sp.	DBH	Cond.	Mgt. Needs		Sp.	DBH	Cond.	Mgt. Needs
037	14	1	0		019	22	2	1
022	16	2	1					
001	36	3	2, 3			ETC.		
011	32	4	7 (3)					
Blank								
	ETC.							

Planting needs _____   Planting needs _____

In this example, we worked north on Pine Street. The west side of the street is on the left of the tally sheet and the east on the right. The first tree on the left side is a red oak, 14 in. in diameter, in good condition with no present management needs. The second is a honeylocust, 16 in., in fair condition in need of minor pruning. The third is a green ash, 36 in., in poor condition in need of major pruning and wound repair. The fourth is an American elm, 32 in., dead or dying, in need of removal. Removal is difficult because of wires and other obstructions. There is no need for replanting after removal. The fifth tree space is blank and should be planted with a potentially medium-sized tree, for example.

With this system, the computer can locate a particular tree only as being in a particular block and street side. For a more precise location, one would have to go to the tally sheets and count down. Distances from intersecting curb lines at the end of the block could be recorded, but practical use of such precise data might not justify the cost.

APPENDIX 4

Sample Format for a Call for Bids in City Tree Maintenance Contracts*

Organization
City of _____ **Date** _____
Call for Bids

Bidders will be required to carefully examine the site of the proposed work and to judge for themselves as to the nature of the work to be done and the general conditions relative thereto; and the submission of a proposal hereunder shall be considered prima-facie evidence that the bidder has made the necessary investigation and is satisfied with respect to the conditions to be encountered, the character, quantity, and quality of the work to be performed.

Each bid shall be in a sealed envelope submitted to the _____ , and will be publicly opened and read by the Park Superintendent or his duly appointed officer at 2:00 P.M., on _____. All proposals presented shall show the proposed prices clearly and legibly and must be properly signed by the bidder, giving address and telephone number.

Bidders must be thoroughly competent and capable of satisfactorily performing the work covered by the proposal and, when requested, shall furnish such statements relative to previous experience on similar work, the plan or procedure proposed and the organization and equipment available for the contemplated work, and any other as may be deemed necessary by the Park Superintendent in determining such competence and capability.

The award of the contract will be to the lowest responsible and qualified bidder whose proposal complies with all the prescribed requirements; but until an award is made, the right will be reserved to reject any or all bids, if to do so is deemed to best serve the interests of the City. In no event will an award be made until all necessary investigations are made as to the responsibility and qualifications of the bidder to whom it is proposed to make such award.

It shall be understood that the contractor will be required to perform and complete the proposed work in a thorough and workmanlike manner, and to furnish and provide in connection therewith all necessary labor, tools, implements, equipment, materials, and supplies except such thereof as may otherwise be specified will be furnished by the City. The City reserves the

---

*This appendix from C. J. Plinkerton, "Contracting of Tree Work and Cost Accounting." *International Shade Tree Conference Proceedings*, Western Chapter, 44:284–293, 1968.

right in the case of a cash contract to make such increase or decrease in the quantity of any item of work to be performed or furnished under such contract; and in the event that any such increase or decrease in the quantity of work to be performed or furnished is so ordered, the amount to be paid the contractor under his or her contract shall be correspondingly increased or decreased, as the case may be, in proportion to the increased or decreased quantities of work.

At the end of each working day the contractor shall clean the site thereof, and all grounds that he has occupied in connection therewith, of all rubbish and debris caused by him; and all parts of the work shall be left in a neat, orderly, and presentable condition. All rubbish and debris as a result of this contract may be disposed of, free of charge, at the Municipal Dump.

All work to be performed shall be subject to the direct supervision of the Park Superintendent and his duly appointed officer, and in all respects shall meet with his approval as conforming with the provisions and requirements prescribed therefor.

Public liability insurance in the amount of not less than $100,000.00 for injuries to or death of any one person, and in the amount of not less than $300,000.00 for injuries to or death of any two or more persons in any one accident and property damage insurance in an amount of not less than $25,000.00 shall be taken out and maintained by the contractor for the duration of the contract. The contractor shall furnish the Park Superintendent with a certificate containing a ten-day cancellation notice clause.

The contractor shall secure, maintain in full force and effect, and bear the cost of, complete Workman's Compensation Insurance for the entire duration of the contract, and the City will not be responsible for any claims, or suits in law or equity, occasioned by the failure of the contractor to do so.

When the work is completed and ready for final inspection, the contractor shall so notify the Park Superintendent. As soon as possible thereafter the Park Superintendent or his duly appointed officer will make the necessary inspection and if he finds that the work has been properly performed and completed in accordance with all terms of the specifications and contract, he will accept it and notify the finance officer to that effect.

During the progress of the work adequate provisions shall be made by the contractor to so accommodate the normal traffic over the public streets as to cause a minimum of inconvenience to the general public. Means of ingress and egress for occupants of property adjacent to the work, with convenient access to driveway, housing, or building shall be provided as far as possible.

The contractor will be required to provide and maintain barriers, guards and lights when and where it may be necessary so to do in order to effectively guard the public from danger as a result of the work being done. He will also be required to post proper notices and signals to the public regarding detours and the condition of the work under construction, all in accordance with applicable provisions of the Vehicle Code.

The contractor will be held responsible for the preservation of all public and private property along and adjacent to the work being done, and will be required to exercise due precaution to avoid and prevent any damage or injury thereto as a consequence of his operation. All trees, shrubs, ground covers, fences, warning signals, and street signs shall be adequately protected and should not be removed or disturbed without permission from the Park Superintendent.

Should any direct or indirect damage or injury result to any public or private property by or on account of any act, omission, neglect or misconduct in the execution of the work, or as a consequence of the nonexecution thereof, on the part of the contractor or any of his employees or agents, such property shall be restored, by and at the expense of the contractor, to a condition equivalent to that existing before the damage or injury occurred, by repairing or rebuilding the same, or by otherwise making good such damage or injury, in an acceptable manner.

APPENDIX 5

Conversion Factors

English to Metric

To Convert from	to	Multiply by
Inches	centimeters	2.540
Sixteenths of an inch	millimeters	1.587
Eighths of an inch	millimeters	3.175
Fifths of an inch	millimeters	5.080
Fourths of an inch	millimeters	6.350
Thirds of an inch	millimeters	8.467
Halves of an inch	millimeters	12.70
Feet	meters	0.3048
Number per bushel	number per hectoliter	2.838
Pounds per bushel	kilograms per hectoliter	1.287
Ounce per bushel	grams per hectoliter	80.44
number per pound	number per kilogram	2.205
Number per pound	number per gram	0.002205
Number per ounce	number per gram	0.3527
Number per square foot	number square meter	10.76
Number per linear foot	number per linear meter	3.281
Degrees, Fahrenheit (F)	degrees, Centigrade (C)	0.55 (F − 32)

Metric to English

To Convert from	to	Multiply by
Millimeters	1/16 of an inch	0.6301
Millimeters	1/8 of an inch	0.3150
Millimeters	1/5 of an inch	0.1968
Millimeters	1/4 of an inch	0.1574
Millimeters	1/3 of an inch	0.1181
Millimeters	1/2 of an inch	0.07874
Millimeters	inch	0.03937
Centimeters	inch	0.3937
Meters	feet	3.281
Number per hectoliter	number per bushel	0.3524
Kilograms per hectoliter	pounds per bushel	0.777
Grams per hectoliter	ounces per bushel	0.0124
Number per kilogram	number per pound	0.4536
Number per gram	number per pound	453.6
Number per gram	number per ounce	28.35
Number per square meter	number per square foot	0.0929
Number per linear meter	number per linear foot	0.8048
Degree Centigrade (C)	degrees, Fahrenheit (F)	(1.8 × C) + 32

INDEX

American Forestry Association, 10, 11, 249
American Society of Consulting Arborists, 195
Arbor Day, 5, 229, 249
Arbor Day Foundation, 249
Arboretum, 4
Arboriculture, 5
  International Society of, 5, 9, 195, 201, 203, 204, 246, 247
  specifications and standards of practice, 282–287
Arborists:
  associations, 141, 195, 203, 237, 244, 247
  certification of, 237, 239
  commercial, 195
  municipal, 6, 271, 287
  training of, 230–239

Benomyl, 235
Boulevards, 3, 19

Casualty losses:
  definition of, 194
  income tax deductions for, 193, 195–200
  insurance, 194
Certification:
  of arborists, 237–239
  of pesticide operators, 234
Citizens Advisory Committee on
    Recreation and Natural Beauty, 6
City tree boards or commissions, 209, 243
  advisory, 209, 210
  operational, 209
  policymaking, 209, 212–213
Civilian Conservation Corps, 5
Commission on Education in Agriculture
    and Natural Resources, 6
Communication, 228–229
Composition:
  control of, 33
  as influenced by:
    mobility, 34
    nostalgia, 35, 40
    physical elements, 30
    purpose, 32
    socioeconomic factors, 33, 40
  as reflected by species popularity, 32, 46

Contracts, bid specifications, 171–173, 291–293
Cook County Forest Preserves, 21
Cooperative Forest Management Act, 7, 242

Disease:
  Dutch elm, see Dutch elm disease
  oak wilt, 6
  phloem necrosis, 5
  white pine blister rust, 4
Dutch elm disease, 5, 32, 39, 40, 43, 45, 46, 122, 124, 144, 147, 167, 174, 187, 235, 242
  control systems, 235
  description of, 167

Energy:
  implications for urban forestry, 258
  potential of urban forest materials, 177–179
Environmental Protection Agency, 235, 239
Erosion:
  control of, 68
  types, 67

Federal Environmental Pesticide Control
    Act of 1972, 234
Financing sources, 186, 215–217, 218–219
Forest Service USDA, 7, 10, 11, 242, 253
  authority for research, 253
  Cooperative Forest Management Act, 7, 242
Formulas, shade tree evaluation, 202–203
  Felt, 202
  Felt-Spicer, 202
  ISA, 203
  ISTC, 203
  Roth, 202
  Stone, 202

Gardens, botanic, 2, 3
Glare:
  control of, 89
  types, 88
Greenbelts, 14, 21

Highway:
  entrances, 33
  rights of way, 19

Information and education:
  methods, 228
  reasons for, 226, 227
Insects, management, 166
  elm bark beetle, 167
  elm leaf beetle, 46, 227
  gypsy moth, 166, 169
Integrated pest management, 167
Internal Revenue Service, 193, 194, 195
International Society of Arboriculture, 5,
    9, 195, 201, 203, 204, 246, 247
Inventories, 140, 141–144, 288–290

Kansas Arborists Association, 210, 237,
    244
Kansas Community Forestry Program,
    242–244

Landfills, 21
Landscape Architects, 4, 229, 244
  architecture, 5, 150
  design elements, 153–159
  design factors, 99–102, 120, 122, 150

Maintenance, 160–173
  contracts, 170
  damage control, 166
  fire control, 169
  growth control, 160–166
  insect and disease management, 166
Management:
  reasons for, 138–140
  rights and responsibilities, 25, 138
Microclimate:
  classes, 55, 125
  factors, 124

National Arbor Day Foundation, 249
National Arborists Association, 203, 246
National Park Service, 5
National Urban and Community Forestry
    Leaders Council, 249
National Urban Forestry Conference, 9,
    250
Noise abatement, role of plants in, 74–85

Occupational Safety and Health Act, 234
Ordinances, 27, 33, 139, 214, 215, 265–287
  elements, 214
  necessity of, 215
  sample, 265–287
  types, 214

Pinchot Institute of Environmental
    Studies, 7
Planning, 221
Planting, 33, 34, 35, 46, 120, 145, 148, 150
  in business districts, 148
  design, 150–160
  location, 145
  species composition control, 144
Pollutants, herbicides, 129
  carbon monoxide, 86
  hydrogen flouride, 85
  manufactured gas, 133
  nitric oxide, 86
  nitrogen dioxide, 85
  ozone, 86, 129, 130
  particulate matter, 87
  rock salt, 131
  sulfur dioxide, 85, 86, 128, 129
Pollution:
  air, 85–88, 128–130
  definition of, 128
  light, 133, 134
  noise, 74
  soil, 130
  water, 69
Property:
  rights, 25
  values as influenced by trees, 192
Pruning, 162
  improper, 162, 163
  priorities, 162
  reasons for, 162
  utility lines, 163–165
Public relations, 227, 228

Removal, 174–180
Research:
  agencies, 252
  needs, 252
  programs, USDA Forest Service, 253
Rights of way:
  easements, 25

highway, 19, 21
railroad, 21
street, 19
Riparian areas, 23, 24

Snow drifting, control of, 61, 63
Society of American Foresters, 8, 247
   Urban Forestry Working Group, 8, 247
Soils, 36, 67, 69, 73, 122
Solar radiation, 50, 88
Sound properties, 75
Space, as factor in planting design, 120, 150, 151, 156
Streetside, trees, 19, 120, 150–151
   area, 14
   design, 150–160
   rights of way, 19
   situations, 145–150

Temperature modification, 50–55
Topography, 123, 146
Traffic control, use of plants for, 91–95
Training:
   methods, 231
   reasons for, 230
Tree City USA, 229, 249
Tree forms, 122
   as factor in planting design, 155
Treelawns, 19, 27, 145, 148, 149
Tree size classes, 120, 156

Urban forest:
   administration, 208–222
   area, 14, 17, 25
   benefits, 51–115
   classification, 140
   composition, 30–36
   definition, 15
   distribution, 14
   example, Manhattan, Kansas, 27, 36–47
   inventories, 141–144, 288–290
   management, 5, 139–180
   ownership, 17–25
   valuation, 184–284
"Urban Forest, The" (movie), 245
Urban forest materials:
   fuel value, 177, 178

harvesting theory, 175
   problems, 175, 179
   removal, 174
   types, 179
   utilization, 179
   volume, 176
Urban forestry:
   concept, 5
   consultants, 245
   definition, 8
   education, 262–265
   history, 2–11
   issues, 258–260
   planning, 221
   program financing, 215–217
   research, 252–254
Utility:
   easements, 25
   pruning, 163–165

Values:
   alternative, 184
   amenity, 158
   city assets, 184
   formulas for, 202–203
   legal, 193
   maintenance, 187
   property, 192
   timber, 187
Volunteers:
   organizations, 251
   use of, 252

Wastewater, treatment and disposal systems, 69–74
Watersheds:
   management, 69
   protection, 67
Wildlife:
   abundance, 103
   hazards, 108
   planning, 109–112
   problems, 107
   values, 107
Windbreaks, 58, 59–63

Zones, plant hardiness, 30, 31